RECENT DISCOVERIES IN
INORGANIC CHEMISTRY

T0296581

RECENT DISCOVERIES IN INORGANIC CHEMISTRY

J. HART-SMITH, A.R.C.S., F.I.C.

CAMBRIDGE

AT THE UNIVERSITY PRESS

1919

CAMBRIDGE
UNIVERSITY PRESS

University Printing House, Cambridge CB2 8BS, United Kingdom

Cambridge University Press is part of the University of Cambridge.

It furthers the University's mission by disseminating knowledge in the pursuit of education, learning and research at the highest international levels of excellence.

www.cambridge.org
Information on this title: www.cambridge.org/9781316633366

© Cambridge University Press 1919

First published 1919
First paperback edition 2016

A catalogue record for this publication is available from the British Library

ISBN 978-1-316-63336-6 Paperback

PREFACE

In the following pages an attempt has been made to give some account of the more important discoveries in Inorganic Chemistry within the last fifteen years so far as they concern the subject as usually taught in Schools. The book is in no sense intended to be a text-book but is rather to be regarded as a supplement to existing text-books. It is evident that in a book written with this object it is impossible to avoid a certain amount of disconnected matter. The book contains particulars of all work coming within its scope that has been published up to the end of last year (1917). It had been intended to add a chapter on radio-activity but pressure of war work has prevented this being included in the present edition.

For some years past the general tendency of inorganic chemistry has had a philosophic basis and more importance is to be attached to efforts to discover the principles under-lying the subject than to attempts to prepare new compounds. Much research work is being done in connection with radio-activity, atomic disintegration, evolution of the elements, line spectra, active nitrogen, the intra-stellar elements and atomic-weight variation.

With regard to atomic-weight variation it is interesting to call to mind that many years ago, before radio-active phenomena were known, Sir W. Crookes put forward the suggestion that there might be more than one inhabitant of each gap in the periodic table. This suggestion, made to explain certain facts then under discussion, was forgotten. At present lead is the only element known to have a variable atomic weight.

Perhaps the most interesting work recently published is that relating to the possible evolution of the elements from hydrogen and helium. Two predictions, based on this assumption, were made and both have proved to be correct. All this work is at present more or less disconnected, and chemists now await a new generalisation that will coordinate these discoveries just as the theories of Dalton and Avogadro coordinated the discoveries of the first part of last century.

The material for much of the work described in this book has been obtained from the publications of the Chemical Society, while the article on combustion is largely based on Professor Bone's report to the British Association, for permission to use which I have much pleasure in thanking the Council of the Chemical Society and Professor Bone.

J. H.-S.

London,
May 18, 1918.

TABLE OF CONTENTS

TABLE OF ABBREVIATIONS EMPLOYED
IN THE REFERENCES

Amer. Chem. J.	American Chemical Journal.
Anal. Fis. Quim.	Anales de la Sociedad Española de Física y Química.
Analyst	The Analyst.
Annalen	Justus Liebig's Annalen der Chemie.
Ann. Chim. Phys.	Annales de Chimie et de Physique.
Ann. des Mines	Annales des Mines.
Ann. Physik	Annalen der Physik.
Ann. Report Chem. Soc.	Annual Reports of the Chemical Society.
Atti R. Accad. Lincei	Atti della Reale Accademia dei Lincei.
Ber.	Berichte der Deutschen chemischen Gesellschaft.
Bull. Soc. Chim.	Bulletin de la Société chimique de France.
Chem. Zentr.	Chemisches Zentralblatt.
Compt. rend.	Comptes rendus hebdomadaires des Séances de l'Académie des Sciences.
Gazzetta	Gazzetta chimica italiana.
J. Amer. Chem. Soc.	Journal of the American Chemical Society.
J. Chim. phys.	Journal de Chimie physique.
J. Pharm. Chim.	Journal de Pharmacie et de Chimie.
J. Physical Chem.	Journal of Physical Chemistry.
J. Physique	Journal de Physique.
J. pr. Chem.	Journal für praktische Chemie.
J. Russ. Phys. Chem. Soc.	Journal of the Physical and Chemical Society of Russia.
J. Soc. Chem. Ind.	Journal of the Society of Chemical Industry.
Monatsh.	Monatshefte für Chemie und verwandte Theile anderer Wissenschaften.
Proc. Amer. Acad.	Proceedings of the American Academy.
Proc. Chem. Soc.	Proceedings of the Chemical Society.
Phil. Mag.	Philosophical Magazine (The London, Edinburgh and Dublin).
Phil. Trans.	Philosophical Transactions of the Royal Society of London.

Proc.K.Akad.Wetensch. Amsterdam	Koninklijke Akademie van Wetenschappen te Amsterdam. Proceedings (English version).
Proc. Roy. Soc. . . .	Proceedings of the Royal Society.
Proc. Roy. Soc. Edin. .	Proceedings of the Royal Society of Edinburgh.
Rec. trav. chim. . . .	Recueil des travaux chimiques des Pays-Bas et de la Belgique.
Sitzungsber. K. Akad. Wiss. Berlin	Sitzungsberichte der Königlich Preussischen Akademie der Wissenschaften zu Berlin.
Trans. Chem. Soc. . .	Transactions of the Chemical Society.
Trans. Faraday Soc. .	Transactions of the Faraday Society.
Verh. Ges. deut. Naturforsch. Ärzte	Verhandlungen der Gesellschaft deutscher Naturforscher und Ärzte.
Zeitsch. angew. Chem. .	Zeitschrift für angewandte Chemie.
Zeitsch. anorg. Chem. .	Zeitschrift für anorganische und allgemeine Chemie.
Zeitsch. Chem. Ind. Kolloide	Zeitschrift für Chemie und Industrie der Kolloide. (Now Kolloid-Zeitschrift.)
Zeitsch. Elektrochem. .	Zeitschrift für Elektrochemie.
Zeitsch. physikal. Chem.	Zeitschrift für physikalische Chemie, Stöchiometrie und Verwandtschaftslehre.

GENERAL

ELEMENTS AND ATOMS

THE discoveries of the present century such as radio-active disintegration and isotopy[1] have brought to light many facts which are incompatible with the ideas attached to the terms element and atom during the last century. The problem of the isotopic elements is not only of immediate interest in relation to the periodic system but has a bearing on some of the fundamental ideas upon which modern chemical theory is based.

In mass action effects isotopes are mutually replaceable and the tests which ordinarily serve for the recognition of chemical individuality lead therefore to the conclusion that isotopes are varieties of one and the same element. Elements may be defined as substances which have not been simplified by any chemical process, while atoms are the ultimate particles which represent the limit of chemical subdivision and which are unchanged by chemical processes.

The known facts relating to elements and their isotopic forms may be summarised thus: (1) the number of the elements is ninety-two, four of which are as yet unknown; (2) each of these elements is characterised by definite chemical and electrochemical properties, by its ordinary and X-ray spectra, and by its nuclear charge; (3) the atomic weight and radio-active properties are not characteristic constants; (4) the number of varieties of atoms is greater than the number of elements, and up to the present some one hundred and twenty varieties have been identified; (5) an element may be pure or mixed; the constituent atoms of a pure element are similar, whereas a

[1] See page 47.

mixed element contains two or more kinds of atoms which differ in weight or in radio-active properties or possibly in respect of both these qualities ; (6) in chemical reactions, the active mass of mixed elements is represented by the sum of the active masses of the different varieties of atoms.

In regard to the views summarised in the above statements, it may possibly be objected that the conception of a mixed element is contrary to the fundamental idea implied by the term element, but against this it need only be pointed out that the new conception, like the old, rests on a purely experimental basis. Whether an element is pure or mixed, it represents the ultimate limit of chemical analysis.

ATOMIC AND MOLECULAR STRUCTURE

An electrical theory of the structure of the atom was put forward by Sir J. J. Thomson[1] in 1904. He assumed the atom to consist of a large number of negative electrons arranged in concentric shells in a sphere of uniform positive electrification. The characteristics of such systems containing different numbers of corpuscles offer a very close analogy to the variation in the valency of atoms in the periodic system with ascending atomic weights, as the hypothetical electrical atoms group themselves with regard to stability in a periodic fashion with regular variations of electropositive and electronegative valencies as the number of corpuscles increases. The non-valent elements find a natural place in the system and the breaking up of the atoms of radio-active elements, with great liberation of energy, finds a satisfactory analogue in electrical atoms of many corpuscles arranged in the shells in groups.

The mass of the negative electron is $\frac{1}{1700}$ of that of an atom of hydrogen, so if the positive electron be ignored

---

[1] *Phil. Mag.* 1904, [VI] **7**, 237.

the atom of hydrogen will contain about 1700 corpuscules; an atom of oxygen about $16 \times 1700 = 27000$ etc. There are objections to the hypothesis, but nevertheless it throws an interesting light on the possible nature of the atom.

In 1911 Rutherford[1] brought forward a theory of atomic structure to account for the large angles through which a very small proportion of the α-particles are deflected, in their passage through matter, by rare exceptionally close single encounters with atoms. On this theory the mass of the atom, in association with positive electricity, about one unit of charge per two units of mass, occupies a single central nucleus of excessively minute dimensions, in diameter only one ten-thousandth of the atomic diameter. Around this nucleus a number of negative electrons, equal to the value of the positive nuclear charge, circulate as an outer ring or shell. This atom is, of course, not stable according to ordinary electrodynamical laws, for nothing apparently operates to prevent the dispersion of the extremely concentrated central positive charge, but it is now recognised that these laws require modification.

It appears that the magnitude of this central positive charge, in terms of the unit atomic charge, for example, that carried by the hydrogen ion, is probably the same as what is conveniently termed the "atomic number[2]." The atomic number is the number of the place an element occupies in the periodic table, when the successive places from hydrogen to uranium are numbered in sequence, hydrogen being unity, helium two, lithium three, and so on. If the eleven known representatives of the rare-earth group, between cerium and tantalum exclusive, are all that exist, the atomic number for uranium would be 89.

During the year 1917 an interesting contribution to

---

[1] *Phil. Mag.* 1911, [vi] **21**, 669.

[2] A. van den Broek, *Nature* 1913, **93**, 373, 476.   F. Sody, *ibid.* 399, 452.   E. Rutherford, *ibid.* 423.

this subject was made by W. D. Harkins in a series of papers on the evolution of the elements from hydrogen[1].

The assumption that the elements are derived from hydrogen by a process in which the formation of the helium nuclei represents a primary and distinct stage in the process of agglomeration leads to the view that the elements, excluding hydrogen, should fall into two series one of these beginning with helium and the other with lithium. Among the elements of low atomic weight the atoms having even atomic numbers are in general built up of helium atoms and have the general formula nHe while those having odd atomic numbers have the general formula $nHe - H_3$, these formulæ representing intra-atomic and not chemical compounds.

Based on this hydrogen-helium-structure hypothesis two predictions were made. The first prediction was that the elements of low atomic number would be found to show evidences in their atomic weights that their atoms are built up according to the general plan, in relation to which the radio-active elements, which are of high atomic weight, disintegrate. The second prediction was that the elements of even atomic number would show a marked difference in abundance from those of odd atomic number. Both these predictions have been verified in a very striking way. Whether the relative abundance in the earth's crust, in meteorites, or in the lithosphere, is considered, the even numbered elements occur in much greater quantity. The abundance of the elements in the earth's crust might seem to give the best information in this respect if it were not known that the surface of the earth has been subjected to very long-continued differentiative processes, and so has a very local character. The meteorites, on the other hand, come from much more varied positions in space, and at the same time show much less indication of differentiation. In

[1] *J. Amer. Chem. Soc.* 1915, **37**, 1367, 1383, 1396; *ibid.* 1916, **38**, 169; *ibid.* 1917, **39**, 856.

the meteorites, the elements of even atomic numbers are, on the average, about seventy times more abundant than the odd-numbered elements, and, moreover, if the elements are plotted in the order of their atomic numbers, it is found that the even-numbered elements are in every case very much more abundant than the adjacent odd-numbered elements. Almost more striking than this is the fact that the first seven elements, in the order of their abundance, are all even-numbered, and, furthermore, make up 98·78 per cent. of the material. Both the iron and the stone meteorites separately show the same relations. Thus the stone meteorites contain 97·6 per cent. and the iron meteorites 99·2 per cent. of even-numbered elements. It is remarkable that the highest percentage found for any odd-numbered element in any class of meteorites is 1·53, whilst among the even-numbered elements larger percentages are common and range even as high as 90·6 per cent. In the lithosphere, whilst the contrast is not so striking, the even-numbered elements are still seven to ten times as abundant as those which are odd, according as the calculations are made by weight or by atomic percentage. Among the rare earths, the even-numbered elements are the more abundant. Among the radio-active elements, the odd-numbered element is in each case either of a shorter period than the even-numbered or else as yet undiscovered. All the five unknown elements are of odd numbers. The elements of low atomic number are found to be much more abundant than those of high atomic number, both in meteorites and on the earth. Thus the first twenty-nine elements make up about 99·9 per cent. of the material, while the remaining sixty-three are either extremely rare or comparatively rare. The variation in the abundance of elements would seem to be the result of an atomic evolution, which is entirely independent of the Mendeléeff periodic system.

PERIODIC LAW

In the first year or so of this century Mendeléeff published his speculations on the chemical composition of the ether. To the groups in the periodic system, in the first place, Mendeléeff proposes to add a zero group in front of group I. In this zero group are placed the elements, helium, neon, argon, krypton, and xenon; a group of elements characterised by their chemical inactivity, for which, therefore, valence is reduced to zero, and further, substances whose molecules are monatomic. Helium belongs to the second series commencing with lithium and ending with fluorine, whilst the first series is represented only by hydrogen, a homologue of lithium, that is, belonging to the same group. The element in the first series of the zero group is represented by "y," a substance which must have the properties characteristic of the argon gases. It is calculated from the relation of the atomic weights of the elements in the neighbouring group that this element has an atomic weight of less than 0·4. The relative density of "y" in relation to hydrogen would be 0·2, and it may be identified with the substance "coronium," whose spectrum was first observed in the corona during the eclipse of 1869. In 1893 some experimenters considered that they had found traces of coronium in their examination of the spectra of volcanic gases.

The molecules of "y" would not be sufficiently light, nor would their velocity be great enough, to identify this element with ether. To complete the series of elements, therefore, a zero series is added, and in this series in the zero group is placed an element "x," which Mendeléeff regards "(1) as the lightest of all the elements, both in density and atomic weight; (2) as the most mobile gas; (3) as the element least prone to enter into combination with other atoms, and (4) as an all permeating and penetrating substance." The element "x," it is suggested, is the ether, the particles and atoms of which are "capable of moving freely everywhere throughout the universe, and have an atomic weight nearly one-millionth that of hydrogen, and travel with a velocity of about 2,250 kilometres per second."

As Mendeléeff points out, "the conception of the ether originates exclusively from the study of phenomena and the need of reducing them to simpler conceptions. Amongst such conceptions we held for a long time the conception of imponderable substances (such as phlogiston, luminous matter, the substance of the positive and negative electricity, heat, etc.), but gradually this has disappeared, and now we can say with certainty that the luminiferous ether, if it be real, is ponderable, although it cannot be weighed, just as air cannot be weighed in air, or water in water. We cannot exclude the ether from any space; it is everywhere and penetrates everything, owing to its extreme lightness and the rapidity of motion of its molecules. Therefore such conceptions as that of the ether remain abstract, or conceptions of the intellect, like that which underlies our teaching about a limited number of chemical elements out of which all substances in nature are composed."

In 1905 Werner[1] suggested an arrangement of the elements in which places could be found for coronium and the ether. The arrangement has many advantages, but difficulties are encountered if all the elements are placed in their order and argon (39·9) has to be placed before potassium (39·2), cobalt (59·0) before nickel (58·7), and tellurium (127·6) before iodine (126·95).

More recent work leading to the electronic theories does not seem to leave any room in the periodic table for the stellar elements. These elements, coronium and nebulium, appear to be of a type of matter altogether unknown upon the earth. (This subject will be more fully dealt with in the chapter on radio-activity.)

[1] *Ber.* 1905, **38**, 914.

ENERGY QUANTA

A theory which has contributed much to the knowledge of specific heats and molecular mechanics has been developed during the last ten years. The principle is generally known as "Planck's quantum law[1]," and may be stated thus: particles of matter emit and absorb energy, not slowly and continuously, but in definite increments which are directly proportional to the vibration frequency of the atom. The increments are called "energy quanta."

The kinetic theory enables the molecular heats of gases to be calculated, and the theory is in entire accord with the facts for monatomic gases where the kinetic energy is entirely due to rectilinear motion. In the case of molecules having two atoms it is assumed that the molecule can rotate in two planes while those having three atoms can rotate in three planes. In these cases the kinetic energy is the sum of the energies of rectilinear motion and that of rotation. The molecular heats of these gases increase with rise of temperature. This discrepancy can be accounted for if it is assumed that as the temperature rises new degrees of freedom appear; for instance, the vibration of an atom about its position of equilibrium. In the case of solid substances the deviations from the law of Dulong and Petit can be accounted for by the theory. The deviation in the case of carbon is shown to be connected with its high melting-point. Lithium, with its low melting-point and small atomic weight, follows the law, as also do lead and mercury, although they have low melting-points and high atomic weights. In the latter case, however, the frequency of vibration of the atom is small compared with that of lithium.

It is found by experiment that at low temperatures[2] the atomic heats of all elementary substances are much lower

---

[1] "Vorlesungen über theoretische Physik," 1910; *Zeitsch. Elektrochem.* 1911, **17**, 265.

[2] E. H. Griffiths and E. Griffiths, *Proc. Roy. Soc.* 1913, [A] **89**, 158.

than at ordinary temperatures and probably vanish entirely at the absolute zero. This diminution in atomic heats with decreasing temperatures can be shown to be a direct consequence of the quantum law. Determinations of the specific heats of elements between the temperatures of boiling nitrogen and boiling hydrogen were made by Dewar in the following way. The substances were cooled in boiling nitrogen and then dropped into liquid hydrogen, a fall of temperature of 57·5°. The volume of hydrogen thus evaporated was measured. The amount of hydrogen evolved from the calorimeter when at rest was well under 10 c.c. per minute, and this was allowed for. As in each experiment the time occupied in the evolution of the hydrogen was about fifteen seconds, this correction only amounted to about 2 c.c. The accuracy of the work is well shown in the case of lead, the mean value of the specific heat of which was found to be 0·02399, the extreme values being 0·0247 and 0·0233. At least three pieces of every substance were used. The results obtained rarely varied by more than 2 to 3 per cent., and frequently they were within 1 per cent. Altogether fifty-three elements were experimented with, about two hundred separate observations being made.

It is found that when atomic specific heats are plotted against atomic weights, a periodic curve is obtained which closely resembles Lothar Meyer's atomic volume curve.

Starting with the assumption that a substance melts when the amplitude of vibration of the atoms about their position of equilibrium is equal to the distance between them, the frequency of vibration can be shown to be

$$2·80 \times 10^{12} \sqrt{T/mv^{\frac{2}{3}}},$$

where T is the melting-point, m the atomic weight, and v the atomic volume at the melting-point. The values for the frequency of vibration obtained by this method are in good agreement with those obtained when the total energy of the atom is calculated according to the quanta theory.

A remarkable confirmation of the quanta theory has been obtained from a consideration of the series spectrum of hydrogen[1]. The hydrogen atom is regarded as a central nucleus charged with an atom of positive electricity and having a negative electron rotating round it. If the motion of this electron is controlled entirely by the laws of Newtonian mechanics no limitation is imposed on the frequency of revolution, and the atom can only emit a continuous spectrum. But hydrogen has a line or series spectrum, and this can only be accounted for by assuming the operation of some principle which restricts the value of the frequency of revolution to definite magnitudes. The quanta theory supplies the necessary principle, and the values of the spectroscopic constants calculated from the theory agree with experimental results.

Further confirmation of the theory is afforded by the newly discovered series of lines belonging to the spectrum of helium[2]. These lines were at first assumed to be due to hydrogen, but it was shown that they could, according to the theory, be given by a helium atom which had lost one of its two electrons. The experiments were then repeated, and it was found that the lines were emitted during the spark discharge through a helium tube which showed no trace of the hydrogen spectrum in any circumstances. From the spectroscopic constants of the series the ratio of the mass of the hydrogen atom to that of the electron can be calculated and has been found to be 1836, a number in extraordinarily good agreement with that given by the best determination of the constant[3].

[1] N. Bohr, *Phil. Mag.* 1913, [VI] **26**, 1, 476; *ibid.* 1914, [VI] **27**, 506.

[2] A. Fowler, *Monthly Notices, Royal Ast. Soc.* 1912, December 73.

[3] A. Fowler, *Phil. Trans.* 1914, [A] **214**, 225.

PHOTOCHEMICAL REACTION

In 1912 a theory[1] was put forward which would seem
to offer an explanation of the mechanism of photochemical
reactions. Every atom in a chemical molecule possesses a
certain amount of free affinity, and the force lines due to
these free affinities must condense together with the escape
of free energy. It therefore follows that the reactivity of
such a condensed system is less than when the system is
unlocked by some suitable means. This unlocking may be
produced by the influence of a solvent or of light or of both
together. Owing to the fact that the light is thereby doing
work against chemical forces the light will be absorbed.
It follows naturally that the chemical reactivity of a
molecular system may be materially increased under the
influence of light of the right wave-length. This theory
would also seem to explain the results obtained by Chapman[2] on the photochemical reactions of chlorine. One of
the most striking results of this work is the inhibiting
action of nitrogen chlorides on the union of hydrogen and
chlorine. It would seem that these substances exert a
negative catalytic action as great as the positive catalytic
action exerted by water on other reactions. Whilst this
action of the nitrogen chlorides might conceivably be due
to their power of absorbing the active light rays, yet it
would seem far more probable that these substances have
the power of preventing the opening of the closed force
systems of the chlorine molecule by light.

A general law for photochemical decomposition on the
basis of the quanta theory has been put forward to the
effect that one quantum of energy brings about the decomposition of a single molecule. This law has been shown to
hold quite accurately for the photochemical production of

[1] *Trans. Chem. Soc.* 1912, **101**, 1469, 1475.
[2] Chapman, *Science Progress*, 1912, **6**, 656, **7**, 66.

ozone from oxygen[1], in which case 46 per cent. of the energy
absorbed is regained when the ozone produced goes back
again to oxygen.

ELECTRICAL RESISTANCE

In 1911 Onnes[2] showed that at 4·3° abs. the resistance
of mercury is only 0·0021 of the value for solid mercury
at 273° abs. The resistance gradually falls between 4·3°
and 4·21°, but between 4·21° and 4·19° it diminishes very
rapidly, and appears to be zero at the latter temperature.
The resistance of platinum also falls very considerably
when cooled from the temperature of liquid helium, but
appears to be constant between 4·3° and 1·5° abs. This
effect is however ascribed to impurities in the metal, and
it is probable that the resistance of pure platinum would
be zero at the temperature of boiling helium. According
to Onnes the disappearance of the resistance of pure metals
at a temperature above the absolute zero is to be antici-
pated according to the quanta theory.

CONSERVATION OF MASS

Many experiments were carried out at the end of last
and during the early part of this century to determine
whether in chemical reactions the total weight of the reacting substances remains quite unchanged or undergoes
appreciable alteration. In one series of experiments carried
out by Landolt[3] the greatest error attaching to an experiment was only 0·03 mgm. As the total mass amounted to
between 300 and 500 grams the experimental error is only
about one part in ten million. Diminution of weight was
found to occur in 61 out of 75 cases. At first it was suggested that the loss might be due to the breaking up of the
atoms on chemical change and to the escape of subatomic

---

[1] E. Warburg, *Sitzungsber. K. Akad. Wiss. Berlin*, 1912, 216.
[2] *Proc. K. Akad. Wetensch. Amsterdam*, 1911, **14**, 113; 1912, **14**, 818.
[3] *Zeit. physikal. Chem.* 1906, **55**, 589.

particles through the walls of the vessel. Landolt however was not convinced that the possibility of systematic errors had been altogether excluded, and decided to re-investigate the problem more fully and to repeat some of the former experiments[1]. Since in most of the reactions investigated heat was evolved, leading therefore to a rise of temperature and a consequent alteration in the amount of water adsorbed on the surface of the glass vessels, experiments were carried out with a view to ascertain the magnitude of this effect, and the length of time required for glass vessels, after being suspended in a desiccator over sulphuric acid, to regain the lost moisture. This was found to be comparatively short, not exceeding two or three days. With regard to the other effects produced by heat, however, these were found to depend not only on the kind of glass, but also on the temperature to which the vessel had been heated ; moreover, the time required for the initial weight of the vessel to be attained after heating varies from about ten to twenty days. It was therefore necessary to take account of the alterations of weight due to these two causes. This was done in respect both of the experiments formerly carried out and of those which were repeated. As a result, it was found that whereas, formerly, 75 per cent. of the reactions seemed to be accompanied by a diminution of weight exceeding the experimental error of 0·03 mgm., now the observed variations of weight were, in almost equal numbers, positive and negative, and all, with four exceptions, were within the experimental error. Even in these four cases the deviations were only slightly greater than this. The final conclusion therefore arrived at was that in the fifteen reactions investigated no change of weight could be detected. Since the total weight employed amounted to about 400 grams, the law of the conservation of mass can be regarded as proved within the limit of accuracy of one part in ten million.

[1] *Zeit. physikal. Chem.* 1908, **66**, 581.

PERIODIC GROUP O

THE gases in this group behave in different ways when brought into contact with charcoal cooled by liquid air and a method of separating neon, krypton and xenon is thereby provided[1]. The same property has been used[2] to determine the proportion of neon and helium in the atmosphere : the percentage of neon has been found to be 0·0000086 by weight and 0·0000123 by volume, while that of helium is 0·00000056 by weight and 0·000004 by volume. The percentage of argon is 0·937.

ARGON

Argon may be prepared from air by causing the latter to pass through a heated iron tube containing calcium carbide which combines with the nitrogen and oxygen[3]. The carbon monoxide is oxydised by copper oxide and the carbon dioxide absorbed by caustic potash, the gas being circulated through the apparatus by a mercury pump.

The compressed oxygen of commerce is now generally obtained from liquid air and contains about 3 per cent. of argon. When the gas is passed over heated copper almost pure argon is obtained : any nitrogen present may be removed by heated magnesium. This is a convenient way of preparing small quantities of argon[4].

---

[1] Valentiner and Schmidt, *Sitzungsber. K. Akad. Wiss. Berlin*, 1905, **38**, 816.
[2] Ramsay, *Proc. Roy. Soc.* 1905, Series A, **76**, 111.
[3] Fischer and Hähnel, *Ber.* 1910, **43**, 1435.
[4] Claude, *Compt. rend.* 1910, **151**, 752.

HELIUM

Several premature announcements of the liquefaction
of helium have been made but the results had been obtained
by using impure material. In 1908 the gas was liquefied
in the following manner[1] : two hundred litres of the gas
were purified by absorption in charcoal at low temperature
in conjunction with chemical methods of eliminating active
elements. Preliminary experiments had shown that the
Joule-Kelvin effect would probably secure condensation
and that therefore the Linde-Hampson method of working
should prove successful. An apparatus similar to that
used for the liquefaction of hydrogen was cooled by considerable quantities of liquid air and hydrogen, and after
three hours work 60 c.c. of liquid helium were obtained.
It is colourless, has a density of 0·15, and boils at 4·5° abs.
while its critical point is below 5·5°. Attempts to freeze
it by rapid evaporation under diminished pressure were
unsuccessful although a temperature only 3° above the
absolute zero was probably reached.

ATOMIC DISINTEGRATION

In 1913 and 1914 some remarkable results on the
formation of neon and helium under the influence of the
electric discharge were published. It had been previously
noticed that many minerals change colour when bombarded
by cathode rays, and that when the gases in the vacuum
tube in which the bombardment had been carried out were
pumped out they were found to contain neon. In 1913
Collie and Patterson[2] independently investigated the phenomenon and subsequently published their results jointly;

[1] Onnes, *Proc. K. Akad. Wetensch. Amsterdam*, 1908, **10**, 744;
11, 168.
[2] *Trans. Chem. Soc.* 1913, **103**, 264.

these appeared to show that both neon and helium were produced in vacuum tubes under the influence of the electric discharge. It has been found that 0·01 c.c. of air contain enough neon to be identified by the spectroscope and so it is possible that an air leak is responsible for the presence of the neon. The experiments were repeated by Merton and Strutt[1] using an apparatus in which the gas was both subjected to the discharge and examined : this avoided a transference to a second piece of apparatus for analysis. After the gas had once been introduced it did not come into contact with any stopcock. The experiments gave negative results. Sir J. J. Thomson[2] also published an account of a long series of experiments carried out on somewhat similar lines. The following results were obtained with the arc discharge in hydrogen at 3 cm. pressure using iron electrodes. After the arc had passed for an hour helium and neon were found. The experiment was repeated on the next day with the same result, on the third day with the same electrodes the presence of helium and neon was doubtful, while on the fourth day the gases could not be detected at all. On using new electrodes the gases were again obtained. This appears to point to the fact that the electrodes were the source of the gases. Exactly similar results were obtained when certain substances were bombarded by cathode rays. It therefore appeared that the presence of neon and helium might be due to (1) air leaks, (2) their presence in the electrodes.

On further consideration however the "air leak" explanation appears impossible. If the presence of neon were due to a small air leak the amount of argon introduced at the same time would be seven hundred times greater and could not possibly escape detection. Again in several cases the formation of helium in greater amounts than the neon was observed, and as the amount of helium present

[1] *Proc. Roy. Soc.* 1914, [A] **89**, 499 ; [A] **90**, 549.
[2] *Nature*, 1913, **90**, 645.

in air is far less than that of neon its introduction as the result of an air leak appears impossible. In 1914 Collie and Patterson[1] repeated their experiments using many different designs of discharge tubes, some with outer jackets. The jackets were either exhausted, filled with neon or filled with water. No difference in the results was obtained so the possibility of contamination due to porosity of the glass of the discharge tube during the discharge is excluded. Collie then used Merton's own apparatus and obtained considerable quantities of helium by the cathode-ray bombardment of powdered uranium metal in an atmosphere of hydrogen[2]. A clue to the cause of the divergent results is given by Collie's observation that when the original 10 in. coil with a platinum break was replaced by a larger coil with a mercury break no helium or neon was produced. If the helium and neon are produced by the cathode-ray bombardment it certainly appears probable that the discharge may be required to be "tuned" to suit the tube used.

The other explanation, the presence of the gases in the electrodes, has also been disproved. Collie and Patterson[3] found that the electrodes were not necessary, for if a powerful oscillating discharge is passed through a coil of wire wound round a bulb containing a little hydrogen, helium and neon are produced. Some mercury was boiled in a vacuum and no trace of either neon or helium was obtained. An electric arc was then passed between electrodes made of this mercury contained in a quartz tube: again neon and helium were formed. One other experiment may be mentioned. Positive results were obtained using a jacketed tube with aluminium electrodes and with hydrogen as the gas. A quantity of the aluminium wire used in making the electrodes was melted in a vacuum

---

[1] *Proc. Roy. Soc.* 1914, [A] **91**, 30.

[2] *Ibid.* 1914, [A] **90**, 554.

[3] *Proc. Chem. Soc.* 1913, **29**, 217.

but gave off no trace of either gas; another portion of the wire was dissolved in an air-free solution of caustic potash, the evolved gas did not contain any neon or helium. The evidence in favour of the assumption that these gases are produced by atomic disintegration appears to be complete.

New Element X$_3$.—Sir J. J. Thomson[1] subjected the gases produced in the vacuum tube to analysis by his positive ray method. In this way he discovered a new gas X$_3$ having an atomic weight of 3. There is no obvious connection between the nature of the gas experimented with and the production of X$_3$, the gas being formed when the vacuum tube was filled with hydrogen, nitrogen, air, helium or a mixture of hydrogen and oxygen. This appears to show that the gas is not triatomic hydrogen. Conditions favourable to the production of X$_3$ generally gave neon and helium also.

[1] *Nature*, 1913, **90**, 645.

PERIODIC GROUP I

HYDROGEN

IN 1904 Travers[1] showed that solid hydrogen is a crystalline substance and not, as stated by previous experimenters, a viscous one.

Calcium hydride[2] when acted upon by water evolves hydrogen, one gram of the solid yielding about one litre of the gas. The substance which is now a commercial product thus provides a convenient source of hydrogen. If aluminium filings are treated with small quantities of potassium cyanide and mercuric chloride, the metal liberates hydrogen when acted upon by water[3]. The prepared metal is placed in a flask and water allowed to drop on to it, the flow of water is regulated so that a temperature of about 70° C. is maintained by the heat of reaction. One gram of the metal gives about 1·3 litres of the gas.

When solutions of metallic salts are heated with hydrogen under pressure the metal is replaced by the hydrogen, the replacement being accompanied by the formation of oxides, hydroxides and basic salts[4]. At 25 atmospheres pressure and at temperature 90° C. a decinormal solution of copper sulphate gives after fifteen hours a basic sulphate, $CuSO_4 . 2Cu(OH)_2$, this then decomposes into cuprous oxide and after 50 hours' treatment metallic copper begins to be formed.

A chemically active form of hydrogen is produced by heating metallic filaments of tungsten, palladium or platinum

---

[1] *Proc. Roy. Soc.* 1904, **73**, 181.
[2] See page 29.
[3] Mauricheau-Beaupré, *Compt. rend.* 1908, **147**, 310.
[4] Ipatieff and Werkhowsky, *Ber.* 1911, **44**, 1755, 3452.

20   HYDROGEN

to high temperatures in hydrogen at very low pressures[1]. The hydrogen at a pressure of 0·01 mm. is allowed to come in contact with the wire at a temperature above 1000° C. The hydrogen appears to dissolve in the metal as atoms, and the atoms diffuse out and owing to the low pressure have little opportunity of joining together again to form molecules. The atoms are absorbed on the glass walls of the apparatus; a better yield is obtained by cooling the walls. This hydrogen can react at room temperature with oxygen, with phosphorus to form the hydride, and with many reducible substances. There are good reasons for thinking that this active hydrogen is hydrogen in the atomic condition.

WATER

In experiments on the behaviour of water under pressure several allotropic forms of ice have been discovered[2]. Ordinary ice is designated Ice I and the newly discovered forms Ice II, Ice III, Ice IV (existence uncertain), Ice V and Ice VI respectively. All these forms except ordinary ice and Ice IV are more dense than water.

Ice II is formed by cooling water first to − 80° C. then compressing to 2700 kg. per sq. cm. and finally cooling with liquid air. It is unstable when the pressure is released crumbling under atmospheric pressure to Ice I at − 130° C. Ice III is prepared by first compressing water to 3000 kg./cm.[2] and then cooling to − 80° C.: the ice so obtained was cooled by liquid air to − 190° C. On removing the pressure Ice III becomes metastable but can be kept in this form sufficiently long to take it out of the pressure vessel and examine it. It is clear and colourless in appearance and differs from Ice I in that it sinks in

[1] Langmuir, *J. Amer. Chem. Soc.* 1912, **36**, 1310; 1914, **36**, 1708; 1915, **37**, 417.
[2] G. Tammann, *Zeitsch. anorg. Chem.* 1909, **63**, 285. *Zeitsch. physikal. Chem.* 1910, **72**, 609.

liquid air. The reversion to Ice I takes place under pressures below 100 kg./cm.^{2} at temperatures from $-120°$ C. upwards. If the transformation is slow the mass becomes opaque and porcelain like, but if the solid be removed from the liquid air the transformation is rapid and the mass falls into a bulky powder melting at $0°$ C.

The formation of Ice IV was deduced from a study of the melting-point curves of ice formed by spontaneous crystallisation under pressures from 500 to 2000 kilograms.

The highest pressure used in the above experiments was 3500 kg./cm.^{2}, but subsequent experiments conducted under pressures up to 20500 kg./cm.^{2} have shown the existence of at least two other forms[1]. The principle of the method employed in the measurements consisted in plotting the displacement of the piston by which pressure is produced against the pressure. The change of phase, being accompanied by a change of volume at constant pressure, is shown by a discontinuity in the curve of piston displacement against pressure. The pressure at the point of discontinuity gives the equilibrium pressure, and the volume swept out by the piston gives the change in volume on passing from one phase to another. The water was placed in a steel shell, open at the top and completely surrounded by kerosene or gasolene, by which the pressure was transmitted. The pressure was measured by observing the change of resistance of a calibrated manganin wire immersed directly in the same chamber with the water, and was controlled at the lower pressures by a Bourdon gauge. The success of the investigation appears to have depended largely on the high efficiency of the piston packing, as no leak occurred even at the highest pressures, although in these circumstances a correction had to be applied for the viscous yield of the steel tubes.

The main results of the investigation are shown in the diagram; the pressures, expressed in 1000 kg./cm.^{2} as unit, are plotted as abscissae against the corresponding

[1] P. W. Bridgman, *Proc. Amer. Acad.* 1912, **47**, 441.

temperatures as ordinates. As usual, each curve represents an equilibrium between two phases, the points where three curves meet—the triple points—equilibrium between three phases.

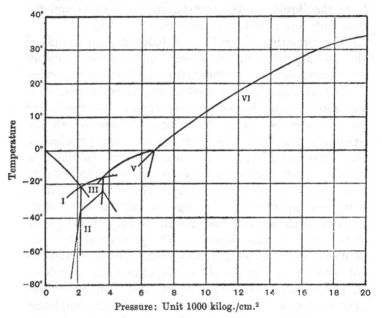

Pressure: Unit 1000 kilog./cm.²

The compressibility of water has also been investigated[1]. It has been found that at low pressures the compressibility at low temperatures is greater than it is at high temperatures, but with rising pressure the abnormality disappears, and above pressures of 4000 kg./cm.² the compressibility is higher at the higher temperatures. Two explanations have been put forward : one assumes that water consists of simple and complex molecules in equilibrium and the amount of polymerisation is diminished by rising pressure, while the other assumes that the double molecules are much more compressible, so the difference in volume between simple and complex rapidly becomes less at higher

[1] P. W. Bridgman, *Proc. Amer. Acad.* 1912, **47**, 446.

pressures. The anomalous decrease of compressibility with rising temperature may be due to the double molecules of high compressibility becoming fewer at higher temperatures. A long column of water appears blue by transmitted light at low temperature but when the temperature is raised it appears green[1]. This can be explained by assuming that polymerised molecules of water are blue while the simple molecules are green. That this is the true explanation is confirmed by the fact that concentrated solutions of colourless salts, which must contain a smaller proportion of polymerised molecules than pure water, are more green than pure water at the same temperature.

LITHIUM

Lithium can be prepared by electrolysing a mixture of its bromide and chloride containing 87 per cent. of the former; the mixture melts at 520° C.[2] A current of 100 amperes at 10 volts is used, a carbon anode and an iron cathode being employed. The metal melts at 180° C.

Up to 1905 there was very little satisfactory information obtainable regarding the basic oxides of the alkali metals, many of the older statements were known to be wrong and much doubt existed as to whether the oxide had ever been prepared in a pure state.

The methods given for the preparation of lithia (combustion of the metal in oxygen, decomposition of the nitrate and the ignition of the carbonate with charcoal) yield impure oxide. The substance can be obtained pure by heating the hydroxide in a current of hydrogen at a temperature of about 680° C. for an hour[3]. Lithium carbonate, which is more easily obtained pure than the hydroxide, can be used, but a temperature of 800° C. is required and the change

[1] J. Duclaux and Mme E. Wollmann, *J. Physique* 1912, [v], **2**, 263.
[2] I. Ruff and Johannsen, *Zeit. Elektrochem.* 1906, **12**, 186.
[3] Robert de Forcrand, *Compt. rend.* 1907, **144**, 1321.

is not complete in less than three hours. The oxide is a colourless, translucent, fused mass which is distinctly volatile below 820° C.

SODIUM, POTASSIUM, RUBIDIUM, CAESIUM

The monoxide of these metals[1] can be obtained pure by partly oxidising the metal in a current of pure dry oxygen and then removing the unchanged metal by prolonged distillation in a vacuum at a moderate temperature. It is probable that a suboxide is first formed which decomposes into metal and oxygen[2].

The oxides are crystalline solids, of density varying from $2\cdot25\,(Na_2O)$ to $4\cdot78\,(Cs_2O)$. Those of sodium and potassium are white when cold and yellow when hot; rubidium oxide is pale yellow when cold and darker when heated; caesium oxide is red. All of them decompose when heated above 400° C. forming the metal and the peroxide. At 200° C. they react with hydrogen forming the hydroxide and hydride

$$M_2O + H_2 = MOH + MH.$$

In the cold they do not react with carbon dioxide or chlorine but do so with greater or less readiness when warmed.

COPPER

The existence of a peroxide of copper appears very probable. If a solution of cupric hydroxide in concentrated sodium hydroxide is electrolysed or sodium hydroxide solution electrolysed alone with copper electrodes a yellow coloured substance is formed at the anode[3]. If hydrogen

---

[1] See also under Lithium.

[2] Etienne Rengade, *Compt. rend.* 1906, **143**, 592, 1152; *ibid.* 1907, **144**, 753.

[3] Erich Müller and Fritz Spitzer, *Zeitsch. Elektrochem.* 1907, **13**, 25.

peroxide is allowed to act on freshly prepared cupric hydroxide at 0° C. a brown crystalline product which loses oxygen on drying is obtained. When acted upon with dilute hydrochloric acid hydrogen peroxide is formed but with concentrated acid chlorine is liberated. The composition of the substance has not yet been determined. It may be Cu_2O_3 or CuO_2.

In 1909 cuprous sulphate was at last prepared pure by the action of methyl sulphate on cuprous oxide at a temperature of 160° C. in the absence of water[1],

$$Cu_2O + (CH_3)_2SO_4 = Cu_2SO_4 + (CH_3)_2O.$$

The cuprous sulphate is washed with perfectly dry ether and dried in a vacuum. When dry the salt is stable·in air but is decomposed rapidly by water with evolution of heat,

$$Cu_2SO_4\,(solid) + (CH_3)_2SO_4$$
$$= Cu_2SO_4\,(solid) + water + CuSO_4\,(dissolved)$$
$$+ Cu\,(solid) + 21\ cals.$$

The development of heat in the reaction shows a difference from the behaviour of other cuprous salts and it explains why the previous attempts to prepare the sulphate in aqueous solution have ended in failure.

From the cryoscopic study of the solutions of copper sulphate, it has been concluded[2] that the molecular complexity of this compound is $(CuSO_4)_2$, and that the composition of the sulphates of bivalent metals in aqueous solutions is represented by the formula $(HMO_4M'')_2O$; the sulphates are supposed to be formed by the condensation of two molecules of sulphuric acid with hydroxides of the formula $(OHM'')_2O$. Many of the reactions of such sulphates, for example, their acidic properties, are readily explained by these assumptions.

---

[1] A. Recoura, *Compt. rend.* 1909, **148**, 1105.

[2] Colson, *Compt. rend.* 1904, **139**, 857.

SILVER

By the action of potassium persulphate on silver pyrophosphate a compound is formed which contains active oxygen and silver in the proportion $1 : 13\cdot12 - 14\cdot89$; it appears to be a derivative of the oxide AgO which requires the ratio $1 : 13\cdot5$[1]. Since it cannot give rise to the formation of hydrogen peroxide and does not reduce lead peroxide, manganese dioxide or potassium permanganate in the presence of concentrated nitric acid, the oxide AgO is not really a peroxide but must be regarded as a basic oxide ; as a base it is weaker than Ag_2O.

When an aqueous solution of silver nitrate is electrolysed between insoluble electrodes, silver is deposited at the cathode and silver peroxynitrate at the anode, there being presumably a simultaneous formation of nitric acid. Both deposits are crystalline, and they grow rapidly towards one another in arborescent crystals. The anode crystals soon become detached, and they are then attacked by the free nitric acid and pass into solution with the liberation of gas. The composition of these crystals was first represented by the formula Ag_7NO_{11}[2]. In 1909 the crystals were shown to contain a true peroxide of silver Ag_3O_4 containing occluded silver nitrate. In 1916 experiments were described in which the electrolyte is kept in continuous circulation, and the decomposing action of the nitric acid is avoided by means of suspended silver carbonate[3]. Two strengths of solution were used, namely, one containing 5 per cent., and the other 20 per cent. of silver nitrate. The percentage of silver in the compound varies from $79\cdot03$ to $79\cdot82$, and the coulometer ratio of the compound to copper deposited in the same circuit varies from $2\cdot98$ to

[1] G. A. Barbieri, *Atti R. Accad. Lincei*, 1907, **16**, 72.

[2] O. Sulc, *Zeitsch. anorg. Chem.* 1909, **61**, 202.

[3] M. J. Brown, *J. Physical Chem.* 1916, **20**, 680.

2·69, but there is apparently no relation between the fluctuations in the coulometer ratio and the silver content. The compound cannot be a pure oxide, because the oxides up to Ag_2O_3 have too high a silver content, and Ag_2O_3 has far too low a coulometer ratio; neither can it be a hydrated oxide of definite composition. The determinations agree with the formula $2Ag_3O_4$, $AgNO_3$, which requires 79·9 per cent. for the silver content and 2·97 for the coulometer ratio. The small differences between the calculated and observed values are probably due to secondary disturbances.

PERIODIC GROUP II

CALCIUM

In 1903 many of the difficulties surrounding the production of metallic calcium by electrolysis were overcome. An iron cathode, so arranged that it can be slowly raised, dips on the surface of the molten chloride. As the metal forms round the end of the rod the latter is slightly withdrawn so that whilst still maintaining connection it gives rise to the gradual production of a rod of the metal calcium. Analysis of a sample of calcium prepared in this way shows it to contain calcium chloride, calcium carbide, iron, silica and manganese, the last named element being possibly derived from the steel tool used to break the calcium[1]. Most of the calcium chloride can be removed by treatment with absolute alcohol[2]. A more efficient purification is effected by fusing the commercial metal with calcium chloride-fluoride to a bright red heat in a closed iron bomb: the product contains about 99·5 per cent. of calcium. The density of the metal obtained in this way is 1·41—1·42.

Calcium may be used for the production of other elements from their chlorides and oxides in "thermite" processes[3]. In some cases fairly pure elements can be obtained but in many the resulting products are alloys. In addition to this latter disadvantage the method suffers from the fact that the lime produced is not fusible at the

[1] Larsin, *Chem. Kentr.* 1905, ii, 1466.

[2] Wilhelm Muthmann, L. Weiss and Josef Metzger, *Annalen*, 1907, **355**, 137.

[3] F. Mollwö Perkin, *Trans. Faraday Soc.* 1907, **3**, 115.

resultant temperature and that the action is very violent. Reactions with the hydride are less violent than those with the metal itself.

Strongly heated calcium is an excellent medium for the separation of the gases of the argon group from mixtures, since all the other gases are completely absorbed by the metal. This power of calcium of absorbing gas has been used as a means of producing high vacua[1]. It appears that there are two forms of metallic calcium, an active and an inactive form[2]. The active modification begins to absorb nitrogen at 300°, and the velocity of the reaction increases with the temperature until it reaches a maximum at 440°, above which temperature the velocity decreases until it vanishes at 800°. The velocity depends on the presence of a layer of the nitride, and only reaches its maximum value after this layer has been formed. The inactive form only commences to combine with nitrogen at 800°. These two forms of metallic calcium do not appear to be allotropic modifications, but merely the metal in two different states of subdivision. When melted calcium is slowly cooled, the active form is produced, and this gives a brown nitride. The inactive form is produced by suddenly cooling calcium from 840°, and it gives a black nitride. The active form absorbs hydrogen between 150° and 300° and above 600° calcium nitride absorbs hydrogen, carbon monoxide, carbon dioxide and methane.

Calcium Hydride. Calcium hydride is manufactured by heating electrolytic calcium contained in horizontal retorts in a current of hydrogen ; the crude product contains 90 per cent. of the hydride. The substance under the name of hydrolite is now a commercial product[3]. When acted upon by water one gram of the substance

[1] F. Soddy, *Proc. Roy. Soc.* 1907, [A] **78**, 429.
[2] A. Sieverts, *Zeitsch. Elektrochem.* 1916, **22**, 15.
[3] Jaubert, *Compt. rend.* 1906, **142**, 788.

gives about a litre of hydrogen which is free from ammonia and acetylene[1].

Calcium Carbide. When nitrogen is passed over calcium carbide under pressure calcium cyanamide, $NC.NCa$, is produced,

$$CaC_2 + N_2 = NC.NCa + C.$$

The crude product known as nitrolim is used as a manure and undergoes decomposition in the soil,

$$NC.NCa + CO_2 + H_2O = NC.NH_2 + CaCO_3$$
$$NC.NH_2 + H_2O = CO(NH_2)_2.$$

The urea thus produced is hydrolysed to ammonium carbonate by certain organisms in the soil. Crude calcium cyanamide is also used for the manufacture of cyanides,

$$NC.NCa + C = Ca(CN)_2.$$

Calcium Hydrogen Carbonate. The maximum quantity of calcium carbonate, which dissolves on shaking for ten hours at 0° in 1 litre of water saturated with carbon dioxide and maintained so at atmospheric pressure, is 1·56 grams, which corresponds with 2·5272 grams of calcium hydrogen carbonate[2]. Under the same conditions at 15° there is dissolved 1·1752 grams of calcium carbonate, which corresponds with 1·9028 grams of calcium hydrogen carbonate. When carbon dioxide is passed very rapidly into lime-water saturated at 15°, the solution finally becomes clear and forms an unstable solution, supersaturated with the gas and containing 2·29 grams of calcium carbonate or 3·71 grams of calcium hydrogen carbonate in 1 litre.

It would appear that calcium hydrogen carbonate can exist in the solid state although it is exceedingly unstable[3]. When a solution of ammonia or potassium hydrogen sulphate is added to a solution of calcium chloride, both being

[1] J. Prats Aymerich, *Anal. Fis. Quim.* 1907, **5**, 173.
[2] A. Cavazzi, *Gazzetta* 1916, **46**, ii, 122.
[3] E. H. Keiser, *J. Amer. Chem. Soc.* 1908, **30**, 1711, 1714.

cooled to zero, a white crystalline precipitate is obtained, the composition of which approximates to that expressed by the formula $CaH_2(CO_3)_2$.

Calcium Sulphate. When gypsum is heated at temperatures not above 210° it becomes anhydrous, but the product is much less dense and more soluble than natural anhydrite[1]. The conversion of the hemihydrate to the soluble anhydrite, which is the last stage in the dehydration process, is reversible, each process only requiring a few minutes at 110° in a dry and a humid atmosphere respectively. Since the hydration to the hemihydrate takes place so easily, this salt naturally forms the principal constituent of plaster of Paris. The best temperature for commercial purposes depends upon the humidity of the air in the oven, thus explaining the variable results that have been obtained. The hemihydrate also absorbs water from the air at the ordinary temperature up to a total water-content of about 8 per cent.

Calcium Sulphide. The cause of the luminosity of the sulphides of the alkaline earths is still unknown, though many attempts have been made to solve the problem. It has been known for a long time that the pure sulphides are not phosphorescent. The phenomenon, when exhibited, only takes place in the presence of an impurity known as the phosphorogen. Polysulphides of the metals must be present though the proportion needs only to be small[2]: the luminosity can be largely increased by the presence of traces of other metals. The conductivity of the powder is increased by exposure to light, it being seven times as great after exposure to a 16 c.p. lamp for an hour as it was in the dark[3]. It has been suggested that the luminosity of the sulphides is due to increased oxidation set up by exposure

[1] G. Gallo, *Gazzetta* 1914, **44**, 1, 497; C. Gaudefroy, *Compt. rend.* 1914, **158**, 2006; **159**, 263.

[2] Vanino and Zumbusch, *J. pr. Chem.* 1911, (ii), **84**, 305.

[3] Vaillant, *Compt. rend.* 1911, **152**, 151.

to light. It has been found, however, that with sulphides which have been sealed up in very pure nitrogen and in as good a vacuum as can be obtained with a mercury pump the power of becoming luminous is maintained unimpaired for three years, during which time traces of oxygen would probably have been absorbed.

The following directions have been given for the preparation of a phosphorescent calcium sulphide[1]: A mixture of 100 parts of calcium carbonate and 30 parts of powdered sulphur is heated at a dull red heat for one hour. It is then cooled and mixed with alcohol, and sufficient of an alcoholic solution of basic bismuth nitrate is added to give 1 part of bismuth to 10,000 parts of calcium sulphide. The mixture is dried in air and then heated at a dull cherry-red heat for two hours, after which it is slowly cooled. The bismuth, as phosphorogen, may be replaced by molybdenum, uranium, or, best, by tungsten.

STRONTIUM

Small quantities of strontium in a crystalline condition have been prepared by heating anhydrous strontium oxide with the calculated quantity of powdered aluminium at a temperature of 1000° C. for four hours[2]. The metal has also been prepared in quantity by the electrolysis of the binary mixture, strontium chloride-potassium chloride[3]. A eutectic point occurs with the mixture containing 15 per cent. of potassium chloride, and when the same method is used as in the case of calcium, sticks of strontium 10 cm. long and 1–2 cm. in diameter can be obtained. The mixture of chlorides melts at 628°, which is 220° below the melting-point of strontium chloride. The current density must be 20–50 amperes per sq. cm. of cathode, and the efficiency is

[1] P. Breteau, *Compt. rend.* 1915, **161**, 732.
[2] Guntz and Galliot, *Compt. rend.* 1910, **151**, 813.
[3] B. Neumann and E. Bergve, *Zeitsch. Elektrochem.* 1914, **20**, 187.

then 80 per cent. Strontium is more oxidisable than calcium, and so soft that it can easily be cut with a knife[1]. It reacts with water, methyl and ethyl alcohols, and aceto-acetic and malonic esters with evolution of hydrogen. Strontium burns in carbon dioxide as readily as it does in air, forming some carbide and free carbon. The hydride is formed by direct union with hydrogen. Analysis of the metal showed that it contained 1·5 per cent. of impurity. The specific heat was found to be 0·0742, and conforms to Dulong and Petit's law, giving 6·5 as the atomic heat.

BARIUM

Barium can be produced by electrolysis by a method similar to that adopted in the case of strontium.

Barium Percarbonate. If carbon dioxide be passed through cold water in which barium peroxide is suspended practically no hydrogen peroxide is liberated till more carbonic anhydride has been used than is equivalent to the barium present[2]. The gas apparently first unites directly with the barium peroxide and forms a yellow solid whose composition is represented by the formula $BaCO_4$. The compound, which has not yet been obtained free from water, is not rapidly decomposed by water, nor does either alcohol or ether remove hydrogen peroxide from it, so it is not merely an additive compound of the latter substance. It can be used for the preparation of hydrogen peroxide by treating it with the appropriate quantity of an acid that forms an insoluble salt.

RADIUM

In 1908 an interesting description was published of the working up of the residues from 30,000 kilograms of pitch-blende residue for radium[3]. The operations extended over

[1] B. L. Glascock, *J. Amer. Chem. Soc.* 1910, **32**, 10.
[2] Wolffenstein and Peltner, *Ber.* 1908, **41**, 275, 280.
[3] L. Haitinger and K. Ulrich, *Monatsh.* 1908, **29**, 486.

two years and the total radium contained in the products was equivalent to rather more than 3 grams of radium chloride.

Metallic radium was prepared by Mme. Curie in 1910[1]. The method was based on the preparation of the amalgam and subsequent expulsion of the mercury. The amalgam was produced by the electrolysis of a solution of radium chloride using a mercury cathode and an anode of platinum-iridium. The weight of radium chloride used was 0·106 gram, and that of the mercury about 10 grams; the amount of radium chloride left in the solution was 0·0085 gram. The amalgam was quite liquid, whereas a barium amalgam prepared under the same conditions contained crystals. Radium amalgam decomposes water, and changes rapidly in air. After drying, it was rapidly transferred to an iron boat, which was heated in a silica tube in hydrogen. The heating must be performed with very great care, since, if the mercury is allowed to boil, some of the radium is lost.

Metallic radium rapidly alters in air, becoming black, probably owing to the formation of a nitride. When a particle of the metal was dropped on to white paper, blackening was produced as if the paper was burnt. When the metal is placed in water energetic decomposition takes place, and most of it dissolves, showing that the hydroxide is soluble, the insoluble residue being probably nitride.

ZINC

Peroxide of zinc has been obtained by the action of 30 per cent. hydrogen peroxide on potassium or sodium zincoxide[2]. It is a white crystalline substance having the composition represented by the formula $ZnO_2 . H_2O$.

[1] Mme Curie and A. Debierne, *Compt. rend.* 1910, **151**, 523.
[2] Kazanecky, *J. Russ. Phys. Chem. Soc.* 1911, **43**, 131.

MERCURY

A peroxide of mercury has been obtained by the action of 30 per cent. hydrogen peroxide on mercury at low temperatures in presence of traces of acid[1]. It is a brick-red powder, which shows no indications of crystalline character[2]. It is decomposed slowly by water, giving basic oxide, oxygen, and hydrogen peroxide; with acids generally, it gives mercuric salt and hydrogen peroxide, but with hydrochloric acid chlorine is formed; with potassium iodide, iodine is liberated; potassium permanganate is decolorised. Unlike lead dioxide, therefore, it is a true peroxide.

[1] Bredig and Antropoff, *Zeitsch. Elektrochem.* 1906, **12**, 585.
[2] Giovanni Pellini, *Atti R. Accad. Lincei*, 1907, **16**, ii, 408.

PERIODIC GROUP III

BORON

CRYSTALLINE boron can be prepared[1] directly by mixing boron trioxide with aluminium turnings and sulphur and starting the action by applying a red-hot iron to a mixture of aluminium and sulphur : the aluminium sulphide is removed from the resultant mass by dissolving it in hydrochloric acid.

Perborates. When boric acid and sodium peroxide are added to water a perborate $Na_2B_4O_8 . 10H_2O$ is obtained[2]. If the compound is treated by hydrochloric acid equivalent to half the sodium contained in it crystals of the perborate $NaBO_3 . 4H_2O$ separate out. The perborates are stable in the solid state, but their solutions, on warming, rapidly decompose with the liberation of oxygen.

ALUMINIUM

When aluminium fluoride is fused with sodium sulphide a mixture of sodium aluminofluoride and sodium thio-aluminate is produced :

$$2Al_2F_6 + 6Na_2S = Al_2F_6 . 6NaF + Al_2S_3 . 3Na_2S.$$

This mixture, when electrolysed in the fused state, yields aluminium, sulphur, and sodium fluoride ; the reactions are expressed as follows :

$$Al_2S_3 . 3Na_2S = 2Al + 3S + 3Na_2S,$$
$$Al_2F_6 . 6NaF + 3Na_2S = 2Al_2 + 3S + 12NaF.$$

On these reactions is based the following patented process: first, aluminium fluoride is formed from bauxite by the action of hydrofluoric acid, the fluoride being freed

---

[1] Kühne, *D. R.-P.* 147871.
[2] Jaubert, *Compt. rend.* 1904, **139**, 711.

from iron and titanium by digestion with alumina. The sulphur produced in the above reactions can be used for the manufacture of the sulphuric acid required to prepare hydrofluoric acid from the sodium fluoride formed, the sodium sulphate resulting from this reaction being converted into sodium sulphide by heating with coal[1].

Aluminium Carbide. It has been found that up to 1400° aluminium carbide acts as a reducing agent on metallic oxides, both constituents being oxidised[2], thus:

$$Al_4C_3 + 12MO = 2Al_2O_3 + 3CO_2 + 12M,$$

whereas at higher temperatures alloys of aluminium and the metal are produced and the carbon only oxidised, for example:

$$3CuO + Al_4C_3 = 4Al . 3Cu + 3CO.$$

These results are explained by the fact that alumina can be reduced by carbon at high temperatures, whereas at low temperatures carbon monoxide oxidises aluminium. The reaction at low temperatures,

$$6Al + 3CO = Al_4C_3 + Al_2O_3,$$

is reversed at higher temperatures.

Rubies and Sapphires. Rubies have been produced[3] by heating alumina with a blowpipe flame into which was blown a mixture of chromium sesquioxide and alumina.

Artificial sapphires were obtained fifty years ago by Deville and Caron, but their results could not be obtained by subsequent workers. In 1910, however, crystals[4], identical in properties with the natural stones, were produced by heating alumina with 1·5 per cent. of magnetic oxide of iron and 0·5 per cent. of titanic acid in the reducing part of the oxyhydrogen flame. The colour is due to the presence of lower oxides of iron and titanium.

[1] Gin, *D. R.-P.* 148627.
[2] Pring, *Trans. Chem. Soc.* 1905, **87**, 1530.
[3] Verneuil, *Ann. Chim. Phys.* 1904, [VIII], **3**, 20.
[4] Verneuil, *Compt. rend.* 1910, **150**, 185.

PERIODIC GROUP IV

CARBON

WHEN Moissan succeeded in preparing artificial diamonds by the process of suddenly cooling the outer portion of a quantity of molten iron containing carbon and then allowing the centre part to cool slowly it was generally assumed that some of the carbon crystallised in the form of diamond owing to the high pressure within the block. Recent experiments[1], however, have rendered this explanation doubtful. Pressures up to 100 tons per sq. inch and temperatures up to the melting-point of magnesia have been employed. The carbon, on cooling, has generally been in. the form of soft graphite and in no case has there been any indication of the production of diamond.

The preparation of a more or less colloidal form of graphite, especially suitable for lubricating purposes[2], has been effected by prolonged agitation of the graphite with a solution of tannic acid ; the material thus obtained passes through filters and remains in suspension for months, but it becomes flocculated on the addition of hydrochloric acid. The method of preparation was based on the known action of tannic acid and other substances on china-clay, by which the latter is rendered much more workable for certain pottery purposes ; a fluid "slip," suitable for casting in forms, can by such means be prepared with much less water than formerly had to be used.

Hydrocarbons. The direct synthesis of methane has been shown to take place at temperatures above $1000°$[3].

---

[1] C. A. Parsons, *Proc. Roy. Soc.* 1907, **79**, [A] 532.

[2] E. G. Acheson, *J. Franklin Inst.* 1907, **164**, 375.

[3] Pring and Hutton, *Trans. Chem. Soc.* 1909, **89**, 1591; *ibid.* 1910, **97**, 498.

At temperatures above 1900° acetylene is also produced and some of this at high temperature decomposes with formation of methane. When small quantities of carbon are used it is possible to convert 95 per cent. of it into methane.

Oxides. The reduction of steam[1] by carbon monoxide takes place between temperatures of 1200° and 1250° and equilibrium is established when the volume of hydrogen is double that of the carbon monoxide as shown in the equation :

$$3CO + 2H_2O = 2CO_2 + 2H_2 + CO.$$

In this reaction traces of formic acid are produced. At temperatures above 1300° carbon dioxide is reduced by hydrogen : the reduction proceeding till one third of the hydrogen has been oxidised:

$$CO_2 + 3H_2 = CO + H_2O + 2H_2.$$

These reactions explain the presence in volcanic gases of the two oxides of carbon, of water and hydrogen and of formic acid.

When malonic acid or its ethereal salts are heated with excess of phosphoric oxide under 12 mm. pressure at a temperature of 300°, various decomposition products are obtained[2]. The products are first cooled[3] to remove unchanged acid, carbon dioxide, ethylene, etc. On passing the remaining substance through a tube cooled in liquid air a white solid with a pungent odour is obtained. The substance melts at $-111°$ C. and boils at 6° C. The composition is represented by the formula C_3O_2, the gas is combustible, burning with a blue smoky flame forming the dioxide. Its formation may be represented by the equation :

$$CH_2(COOH)_2 - 2H_2O = C_3O_2.$$

[1] Gautier, *Compt. rend.* 1906, **142**, 1382.
[2] Diels and Wolf, *Ber.* 1906, **39**, 689.
[3] A. Stock and H. Stoltzenberg, *Ber.* 1917, **50**, 498.

When treated with water it reforms malonic acid. Its
molecular constitution is uncertain, it may be $O:C:C:C:O$
but there is evidence[1] for regarding it as the lactone of
β-hydroxypropiolic acid in which case its structural formula
would be

$$
\begin{array}{c}
\mathrm{C} \\
\diagup\!\!\diagup \\
\mathrm{C} \quad \mathrm{O} \\
\diagdown \diagup \\
\mathrm{C\!:\!O}
\end{array}
$$

The polymerisation of the gas to a red substance is very
readily brought about by the presence of the polymeride
itself. The gas may sometimes be kept for days but as
soon as the polymerisation begins it completely disappears
within a day. In contact with phosphoric oxide the gas
polymerises in less than a minute. For this reason it is
necessary to remove the gas as fast as it is prepared from
contact with the phosphoric oxide.

No satisfactory name has yet been given to this oxide:
the name of carbon suboxide having been applied to compounds discovered in 1873 and 1876. The identity of these
compounds to which formulæ C_4O_3 and C_8O_3 have been
assigned is not well established.

Sulphide. When nickel carbonyl and thiocarbonyl
chloride react or when the vapour of carbon disulphide is
submitted to the silent electric discharge carbon mono-
sulphide is produced[2]. The substance collected in a U-tube,
cooléd by liquid air, was at first white, but even at a
temperature of $-210°$ transformation to a brown substance
takes place in fifteen minutes. The transformation is
accompanied by a flash and sometimes by a detonation
sufficiently violent to shatter the tube. The brown substance, evidently a polymeride, has a composition represented
by the formula $(CS)_n$.

[1] Michael, *Ber.* 1906, **39**, 1915.
[2] Sir J. Dewar and H. O. Jones, *Proc. Roy. Soc.* 1910, [A] **83**, 408;
ibid. 1911, [A] **85**, 574.

Cyanogen. It has been shown that carbon and nitrogen are perfectly pure if both show no tendency to unite when heated with each other even up to the temperature of the electric arc[1]. Previous experiments, in which some cyanogen had been obtained, had probably been made with impure carbon. If pure carbon and nitrogen could unite equilibrium should be obtained in the decomposition of cyanogen, but when the gas is submitted to the action of powerful electric sparks it is completely decomposed.

Percarbonates. Recent work on the percarbonates has shown that it is necessary to differentiate between the true percarbonate and the ordinary carbonate containing hydrogen peroxide of crystallisation[2]. The former give a quantitative separation of iodine from neutral potassium iodide solution, while the latter liberate practically no iodine. Four salts may be prepared by the action of carbon dioxide on sodium peroxide : these are derived from two distinct acids. One acid, peroxycarbonic acid H_2CO_4, gives rise to two salts Na_2CO_3 and $Na_2CO_4 . H_2O_2$, while the other peroxydicarbonic acid $H_2C_2O_6$ forms $Na_2C_2O_6$ and $Na_2C_2O_6 . H_2O_2$.

COMBUSTION

The first exact knowledge of the chemistry of burning was mainly derived from the researches of Davy and his contemporaries (1815—1825). This work was primarily undertaken to elucidate the causes of explosions in coal mines and disclosed the broad facts connected with the ignition of explosive mixtures, the influence of narrow passages and of cold surfaces in extinguishing flames, the relative " combustibilities " and " explosive limits " of inflammable gases and the effects of rarefaction and dilution

[1] Marcellin Berthelot, *Compt. rend.* 1907, **144**, 354.
[2] Riesenfeld and Mau, *Ber.* 1911, **44**, 3589, 3595; *ibid.* 1909, **42**, 4377.

upon gaseous combustion. Davy also discovered the flame-less combustion of hydrogen and coal gas in contact with a glowing spiral of platinum. The importance of the "intensifying" influence of hot surfaces has not been fully appreciated until recently.

The work of Davy was very fruitful in its immediate practical results but gave rise to no great theoretical developments. Soon after his death there arose the mistaken dogma of selective combustion of hydrogen in hydrocarbon flames and this continued to dominate chemical science for more than half a century. It is possible that the doctrine was suggested to Davy's immediate successors by his mistaken views, which are still prevalent, concerning the much higher combustibility of hydrogen as compared with hydrocarbons.

Bunsen's researches upon gaseous combustion introduced more exact methods of gas analysis and elucidated the reducing action of carbon monoxide in the blast furnace. His experiments on the division of oxygen between carbon monoxide and hydrogen are now recognised to have been vitiated by the fact that he worked with undried gases in a "wet": eudiometer. Bunsen put forward the idea of "discontinuity" or variation *per saltum* in regard to gaseous combustion. This notion, although disproved, is still met with. Bunsen's first measurements of the rates at which flames are propagated have since been shown to apply only to the initial stages of an explosion, the final constant velocity of the explosion wave is very much higher.

Ignition Temperatures. If electrolytic gas be heated in a sealed bulb to a temperature just above 400° C. the formation of steam can usually be detected after a few days. Between 450° and 500° the rate of combination though considerably greater would still be insufficient to cause any self-heating of the mixture. If however the temperature of the enclosure be further slowly raised a point (about 550°) would soon be reached at which

self-heating of the mixture would begin ; its temperature
would thus be raised above that of the enclosure and the
rate of combination rapidly accelerated until explosive
combustion would be set up. If the preliminary heating
takes some time the mixture will be diluted by the products
of its own combustion and so the true ignition point will
not be obtained. The difficulty[1] can best be overcome by
heating the combustible gas and oxygen separately to the
ignition point before allowing them to mix. Neglect of
this precaution doubtless accounts for the very discordant
results for ignition temperatures that have been published.

Inflammation and Detonation. During the initial period
of an explosion (inflammation) not only is the flame propagated with a much slower velocity but also the actual
process of combustion is much more prolonged than in
detonation[2]. The rate of inflammation varies considerably
for different mixtures, the following figures being given by
Le Chatelier[3] :

Mixture	Rate
$2CO + O_2$	2 metres per sec.
$2H_2 + O_2$	20 „ „ „
$CS_2 + 3O_2$	22 „ „ „
$2C_2H_2 + 5O_2$	200 „ „ „

The flame sends out invisible compression waves in all
directions, which travel in advance of the flame with velocity of sound through the unburnt gases. Some of these
will be reflected and if the reflected wave is travelling in
the same direction as the flame it accelerates the latter and
causes quickened combustion. The velocity in the flame
is thus continually increased until detonation is set up at
one or more points. In detonation the temperature of each
successive layer of the explosive mixture is suddenly raised

[1] H. B. Dixon and H. F. Coward, *Trans. Chem. Soc.* 1909, **95**, 514.
[2] H. B. Dixon, *Phil. Trans.* 1903, [A] **200**, 315.
[3] Le Chatelier, *Annales des Mines*, 8ᵉ Sér. 4.

to the ignition point by adiabatic compression, and it is probable that a large proportion of collisions between chemically opposite molecules are fruitful of change. The whole combustion is probably completed in an immeasurably short interval of time as the result of a comparatively limited number of successive molecular collisions. There can be little doubt as to the important part played by reflected waves in determining the violent shattering effects associated with gaseous explosions on a large scale.

The Influence of Moisture upon Combustion. Nearly forty years ago H. B. Dixon discovered that a mixture of carbon monoxide and oxygen, dried by long contact with phosphoric anhydride, will not explode when sparked in the usual way in a eudiometer, whereas the presence of a trace of moisture or any gas containing hydrogen (*e.g.* methane, ammonia or hydrochloric acid) at once renders the mixture explosive. The amount of moisture required is extremely small, the presence of 4 molecules of steam per 1000 million molecules of gas being sufficient to allow the action to proceed.

Various theories respecting the function of moisture have been put forward but none is entirely satisfactory. H. B. Dixon[1] maintains that in the combustion of carbon monoxide steam merely acts as a carrier of oxygen

$$CO + OH_2 = CO_2 + H_2.$$

This explanation cannot be extended to all cases and is inapplicable in the case of hydrogen. According to H. E. Armstrong chemical actions cannot occur between two perfectly pure substances but require the conjunction of an electrolyte in order to form a closed conducting system. The pressure of steam, which he supposes may always be regarded as rendered "conducting" by association with some traces of an electrolytic impurity, provides the neces-

---

[1] H. B. Dixon and J. E. Russell, *Trans. Chem. Soc.* 1897, **71**, 605.

sary conditions for the passage of the current, the oxygen
playing the part of depolariser, thus :—

Before				After	
O	H_2O	CO	$\longrightarrow$	OH_2	OCO
O	H_2O	CO		OH_2	OCO

A serious objection to this theory is the fact that there are
several well established cases in which combustion apparently does not depend upon the presence of moisture,
e.g. hydrocarbons, cyanogen and carbon disulphide.

The Combustion of Hydrocarbons. In 1892[1] it was
discovered that an equimolecular mixture of ethylene and
oxygen yields on detonating almost exactly twice its own
volume of carbon monoxide and hydrogen in accordance
with the empirical equation

$$C_2H_4 + O_2 = 2CO + 2H_2.$$

When at about the same time[2] the discovery of hydrogen
in the interconal gases of aerated hydrocarbon flames was
made the dogma of the preferential combustion of hydrogen
became untenable. The idea of selective combustion is
however so opposed to well-established principles that it
could not be expected to meet with general acceptance.

W. A. Bone[3] has made a systematic study of hydrocarbon
combustion at temperatures below the ignition point. The
experiments were conducted in an apparatus in which large
volumes of the reacting mixtures could be circulated at a
uniform speed in a closed system comprising (1) a surface
of porous porcelain maintained at a constant temperature
in a combustion furnace, (2) suitable cooling and washing
arrangements for the removal of condensable or soluble
intermediate products, and (3) a mercurial manometer for
recording pressures.

[1] Lean and Bone, *Trans. Chem. Soc.* 1892, **61**, 873; H. B. Dixon,
Phil. Trans. 1893, 159.
   [2] Smithells and Ingle, *Trans. Chem. Soc.* 1892, **61**, 209,
   [3] *Trans. Chem. Soc.* 1902, **81**, 536; 1903, **83**, 1074; 1904, **85**, 693, 1637.

By means of these experiments it was proved as regards slow combustion (1) that a hydrocarbon is ultimately burnt to a mixture of steam and oxides of carbon without any separation of carbon or liberation of hydrogen at any stage of the process ; (2) that the oxidation is marked by a very large intermediate formation of aldehydic products; (3) that the fastest rates of oxidation are (in the case of the hydrocarbons examined) always obtained with a ratio of hydrocarbon to oxygen between 2:1 and 1:1, an excess of oxygen above an equimolecular ratio always having a marked retarding influence; and (4) that a large proportion of carbon dioxide is often found in the products under conditions which preclude all possibility of its formation either by direct oxidation of the monoxide or by the interaction of the monoxide with steam. The balance of evidence was so great in favour of the supposition that combustion proceeded by successive stages of "hydroxylation" that the following schemes were put forward for the typical hydrocarbons ethane, ethylene and acetylene :

$$CH_3 . CH_3 \rightarrow CH_3 . CH_2OH \rightarrow CH_3 . CH (OH)_2 \rightarrow CO + H_2O + HCHO \rightarrow H . COOH \rightarrow CO (OH)_2$$

Ethane / Ethyl Alcohol — $CH_3 . CH (OH)_2$; $CH_3 . CHO - H_2O$ Acetaldehyde — Formic Acid: Formaldehyde $CO + H_2O$; Carbonic Acid: $CO_2 + H_2O$

$$H_2C : CH_2 \rightarrow H_2C : CH (OH) \rightarrow HO . HC : CH . OH \rightarrow H . COOH \rightarrow CO (OH)_2$$

Ethylene / Vinyl Alcohol ; $2H . CHO$ Formaldehyde ; $CO + H_2O$; $CO_2 + H_2O$

$$HC : CH \rightarrow HO . C : CH \rightarrow HO . C : C . OH \rightarrow H . COOH \rightarrow CO (OH)_2$$

Acetylene ; $CO + H . CHO$; $CO + H_2O$; $CO_2 + H_2O$

In other words the attack of the oxygen upon the hydrocarbon may be supposed to involve a series of successive "hydroxylations," the hydroxylated molecules either breaking down or undergoing further oxidation, according to their relative stability and affinities for oxygen at the particular temperature, substantially as represented by the above schemes.

LEAD

In the extraction of lead from galena the ore is first roasted in air at a low temperature so that lead oxide and lead sulphate are produced[1]. The subsequent changes are somewhat more complex than was first thought. This is due to the fact that reversible reactions are involved and investigation of the equilibrium conditions shows that the following four equations probably represent the chemical reactions that actually occur :

$$PbS + PbSO_4 \rightleftharpoons 2Pb + 2SO_2$$
$$PbS + 3PbSO_4 \rightleftharpoons 4PbO + 4SO_2$$
$$PbS + 2PbO \rightleftharpoons 3Pb + SO_2$$
$$Pb + PbSO_4 \rightleftharpoons 2PbO + SO_2.$$

Lead from Radio-active Minerals. Professor Soddy[2] in 1913 gave an account of a generalisation of radioactivity phenomena which involved the existence in some of the places in the periodic table of several elements, differing in atomic weight, but remarkably similar in other properties. All radio-elements which are chemically identical in character should occupy the same position in the periodic table. Such elements are termed isotopic. The unknown end products of all the known disintegration series fall into the place occupied by lead. The atomic weight of thorium being 232·4, the atomic weight of the thorium isotope should be 208·4, while the atomic weight of uranium being 238·5 that of its isotope should be 206·4.

Ceylon thorite contains 62 per cent. of thorium oxide and only 0·4 per cent. of lead oxide[3]. The very small quantity of lead suggests that it is all of radio-active origin, none being present as an original constituent. The lead was extracted from a kilogram of the mineral and after

[1] Schenck and Rassbach, *Ber.* 1907, **40**, 2185.
[2] *Ann. Report Chem. Soc.* 1913, 262.
[3] F. Soddy and H. Hyman, *Trans. Chem. Soc.* 1914, **105**, 1402.

most careful purification 1·2 grams of lead chloride were obtained. The atomic weight was determined by two different methods, one series giving a value 208·5 and the other 208·4. The atomic weight of lead[1] from N. Carolina uranite has been found to be 206·4, that from Joachimsthal pitchblende 206·6 and from English pitchblende 206·8. The atomic weight of common lead determined in the same way was 207·2.

These very remarkable results undoubtedly support the generalisation as to isotopic elements. A spectroscopic examination of lead of radio-active origin showed no difference between it and ordinary lead. The view that isotopes cannot be separated from one another by chemical means is now generally held and the result of work on the solubility of salts of lead isotopes published last year confirms this[2]. The nitrate obtained from carnotite lead, containing one part of ordinary lead, three parts of radium-G and a trace of radium-B, was submitted to fractional crystallisation more than a thousand times. A determination of the atomic weight of the metal in the least and most soluble fractions gave numbers agreeing within 0·006 per cent., which is well within the limits of the possible experimental error. The β-ray activity of the two fractions was also found to be identical within the limits of the experimental error of 1 per cent. From these observations it is inferred that the molecular solubilities of the nitrates of the isotopes are identical, and further evidence is thus obtained in support of the view that isotopes cannot be separated by crystallisation processes.

If the molecular solubilities of the salts of isotopic elements are identical, then, since the molecular weights are different, it follows that the solubilities, expressed in grams per litre, and the densities of the saturated solutions.

[1] T. W. Richards and M. E. Lembert, *J. Amer. Chem. Soc.* 1914, **36**, 1329.
[2] T. W. Richards and N. F. Hall, *J. Amer. Chem. Soc.* 1917, **39**, 531.

ought to show slight differences. Measurement of the
densities of saturated solutions of the nitrates prepared
from ordinary lead of atomic weight 207·15, and from carnotite lead with an atomic weight of 206·59, has shown that
there is an appreciable difference, the density of ordinary
lead nitrate solution being the greater. Assuming that the
molecular volumes of the isotopes are identical, and that
the change of volume in the formation of the saturated
solutions is the same for both, it would seem that the
difference in the densities of saturated solutions of salts
of isotopic elements may be utilised for the determination
of the relative atomic weights of the isotopes. The difference
between the densities should bear the same ratio to the
mean density as the difference of the atomic weights to the
mean atomic weight. Experiments[1] made with the nitrates
of ordinary, carnotite, and pitchblende lead, have, indeed,
given results which appear to justify the application of the
method.

Very careful investigations[2] have failed to reveal any
difference whatever in the atomic weights of copper, silver,
iron, sodium and chlorine obtained from the most varied
sources possible.

[1] K. Fajans and M. E. Lembert, *Zeitsch. anorg. Chem.* 1916, **95**, 297.
[2] T. W. Richards and M. E. Lembert, *J. Amer. Chem. Soc.* 1914,
36, 1329.

PERIODIC GROUP V

NITROGEN

IF hydrogen be burnt in a carefully regulated stream of air and the products passed over heated copper and copper oxide, "atmospheric nitrogen" is obtained and can be collected in the usual way [1]. This method will be found useful for the preparation of the gas in the laboratory.

The temperature at which nitrogen is absorbed by metals varies greatly [2]: with aluminium, calcium, chromium and magnesium the process begins at about 800°, in other cases no action occurs below 1250°. In only three cases are definite compounds formed, the composition being represented by the formulæ Mg_3N_2, Ca_3N_2 and AlN, in the other cases the product is probably a solid solution of either nitrogen or metallic nitride in the metal. The product obtained when either chromium or titanium is used is distinctly magnetic, while when magnesium is used the product is almost as magnetic as iron. When metals are heated in ammonia instead of nitrogen similar products are obtained [3].

Active Nitrogen. When an electric discharge is passed through a vacuum tube containing small quantities of gas, an after-glow is seen which may last for several minutes. It was supposed that the glow might be due to a recombination of atoms to reform the molecules which had been broken up by the discharge.

In 1911 Strutt [4] brought forward evidence to show that in the case of nitrogen the gas has properties which are not possessed by ordinary nitrogen. A current of nitrogen at

[1] Hulett, *J. Amer. Chem. Soc.* 1905, **27**, 1415.
[2] I. I. Shukoff, *J. Russ. Phys. Chem. Soc.* 1908, **40**, 457.
[3] G. G. Henderson and J. C. Galletly,*J. Soc. Chem. Ind.* 1908, **27**, 387.
[4] R. J. Strutt, *Proc. Roy. Soc.* 1911, [A] **85**, 219; *ibid.* 1911, [A] **86**, 56.

low pressure was drawn through a bulb through which a high tension discharge was being passed. The glowing gas was subjected to various treatment in a tube some 20 cm. further on. Other experimenters using very carefully purified nitrogen failed to obtain the after-glow, and hence concluded that active nitrogen could not be obtained from pure nitrogen. The production of the after-glow, however, should not be regarded as the test for the presence of active nitrogen, as the real test lies in the reaction that can be carried out by its means. It has not yet been absolutely determined if active nitrogen can be prepared from absolutely pure nitrogen. There is no doubt that the form of apparatus and the kind of discharge have some effect and that for the maximum production of the after-glow a catalyst must be present[1]. This catalyst may consist of a trace of oxygen, hydrocarbon gas, carbon dioxide or hydrogen sulphide. Active gas can also be prepared by passing nitrogen at 15 mm. pressure through a direct current arc.

Active nitrogen emits its energy more quickly and reverts sooner to ordinary nitrogen if it is cooled[2]. If the glowing gas is compressed to a small volume it flashes out with great brilliance and loses its activity. The active nitrogen may revert to ordinary nitrogen in two ways: one consists of a volume change accompanied by a glow and the other of a surface action on the walls without glow. Ordinary phosphorus is converted into amorphous phosphorus by the passage over it of the nitrogen. If the gas be passed into the vapour of mercury, cadmium, zinc, arsenic, sodium or sulphur the corresponding nitride is formed[3]. Ethylene and acetylene react vigorously with the gas, the after-glow disappears, and the region where the two gases meet is marked by a lilac flame showing the spectrum of cyanogen:

[1] N. B. Baker, E. Tiede, R. J. Strutt and E. Domcke, *Ber.* 1914, **47**, 2283.

[2] R. J. Strutt, *Proc. Roy. Soc.* 1912, [A] **86**, 262.

[3] *Ibid.* 1913, [A] **88**, 539.

hydrogen cyanide is formed[1]. With pentane, ammonia and amylene and hydrogen cyanide are formed. The action with oxygen is described in the section dealing with the oxides of nitrogen.

Ammonia. The synthetic production of ammonia under the influence of electric sparks has been found to depend upon the temperature[2]. The gaseous mixture of the elements is prepared by passing dry ammonia over electrically heated spirals of nickel, platinum or iron and removing undecomposed ammonia by sulphuric acid. The proportion of ammonia formed, which amounts to only 3—4 per cent. at the ordinary temperature, increases with falling temperature, and the action is nearly complete at the temperature of liquid air. The influence of pressure was also studied, and found to be peculiar, since, within the working range of 20—800 mm., there was found to be (at 100 mm.) a marked maximum in the efficiency with regard to the energy expended; the production at this pressure was 0·17 gram of ammonia per kilowatt hour. The behaviour of ammonia and mixtures of nitrogen and hydrogen at high temperatures has been examined. It is probable that ammonia cannot be synthesised by heat[3], but is produced by the direct union of its elements when the gases are "ionised" by exploding with oxygen, by heating with many metals or by sparking the mixed gases.

Another method for the synthetic production of ammonia was patented in 1907[4]. The process depends on the alternation of the action of hydrogen on calcium nitride, which yields ammonia and calcium hydride, with that of nitrogen on calcium hydride, which also yields ammonia and regenerates calcium nitride; it is advisable that the solid material should be finely divided and expose a large surface. The

[1] A. Koenig and E. Elöd, *Ber.* 1914, **47**, 516.
[2] E. Briner and E. Mettler, *Compt. rend.* 1907, **144**, 694.
[3] Perman, *Proc. Roy. Soc.* 1905, [A] **76**, 167.
[4] Karl Kaiser, *D. R.-P.* 181657.

most suitable temperature is between 200° and 400°, although the reaction sets in about 150°.

Although it was known that ammonium hydroxide existed in small quantity in a solution of ammonia, the substance was not isolated till 1909[1]. In that year a solid, which is probably a definite compound, was obtained having the composition represented by the formula NH_4OH. The freezing-points of solution of ammonia of concentrations from 4·1 to 100 per cent. were determined and the results plotted on a curve which showed maxima at 49 and 65 per cent. The monohydrate should contain 48·6 per cent. of ammonia, and the hydrate, $2NH_3 . H_2O$, 65·4 per cent. The liberation of ammonia from aqueous solutions of ammonium salts heated tó boiling-point is due to the hydrolysis of these compounds and not to dissociation.

Oxides. Nitrogen trioxide[2] has been prepared by the interaction of (i) liquid nitric oxide and liquid oxygen, (ii) gaseous nitric oxide and oxygen at −110° C., (iii) nitrogen peroxide and nitric oxide at −150° C. The trioxide is stable under the ordinary pressure at −21° and by cooling can be converted to a blue solid, melting at 103° C.[3]

It is doubtful if nitrogen pentoxide had been prepared pure before 1913, and hence none of the data concerning its physical constants published before then is correct[4]. The nitrogen pentoxide obtained by the fractional distillation of nitric acid and phosphoric oxide is far from pure. The chief causes of impurities are (i) the tendency of the substance to dissociate into lower oxides of nitrogen, and (ii) the incomplete dehydration of the acid. The tendency to dissociate is prevented by carrying out the distillation in a current of ozonised oxygen, and complete dehydration is obtained by passing the mixed gases through a tube con-

[1] F. F. Rupert, *J. Amer. Chem. Soc.* 1909, **31**, 866.
[2] Francesconi and Sciacca, *Gazzetta*, 1904, **34**, [1], 447.
[3] Wittorf, *Zeit. anorg. Chem.* 1904, **41**, 85.
[4] F. Russ and E. Pokornay, *Monatsh.* 1913, **34**, 1027, 1051.

taining phosphoric oxide. The pure oxide is completely condensed at $-80°$. The vapour pressures have been determined between $-80°$ and $-17·5°$. It is found that above $0°$ the oxide is partly dissociated. As the vapour pressure reaches one atmosphere before the substance melts the oxide has no true melting-point or boiling-point.

Under suitable conditions the pentoxide can be prepared synthetically. By submitting air to the silent electric discharge in presence of water on an alkali, nitric acid or a nitrate may be obtained thus:

$$2N_2 + 5O_2 + 2H_2O + Aq = 4HNO_3 \text{ (dil.)}.$$

It is possible that nitrogen peroxide is first formed and that this is oxidised by the ozone.

When nitric oxide is passed into liquid oxygen a substance called nitrogen hexoxide, NO_3, is produced[1]. Nitrogen hexoxide when suspended in liquid oxygen has a green colour, but as the oxygen boils away the fine flocks unite and a dull greyish-blue compact powder is obtained. The change in colour is accompanied by decomposition, and a loss of oxygen occurs. If the product be washed with liquid nitrogen more oxygen is removed, till finally a dull blue powder of constant composition is obtained. This substance has a formula NO_2 or N_2O_4. It cannot be ordinary nitrogen tetroxide because this at low temperature exists as white crystals. It is concluded that the substance is an isomeride of the tetroxide and so is named nitrogen isotetroxide.

In 1912 Lowry[2] found that if air was passed through an ozoniser or through spark gaps no nitrogen peroxide was produced, but if the air was passed through the spark gaps and ozoniser arranged in series then about 1/4000th part by volume of nitrogen peroxide was produced. The same result was obtained if the air was passed through the spark gaps first and then through the ozoniser, or if the ozoniser and spark gap were used in parallel and the air emerging from

[1] F. Raschig, *Zeitsch. anorg. Chem.* 1913, **84**, 115.
[2] *Trans. Chem. Soc.* 1912, **101**, 1152.

them was mixed. Evidently, therefore, the formation of
nitrogen peroxide cannot be due to a simple electrical action,
but must be the result of a chemical reaction between ozone
and some oxidisable form of nitrogen produced in the spark
gap. It has also been found that when pure oxygen is passed
through an arc, a weak, bluish-green after-glow is formed[1].
If the glowing oxygen is mixed with glowing nitrogen from
a second apparatus both glows are immediately extinguished
and oxides of nitrogen are formed in quantity. If the arc
is extinguished in either apparatus, the formation of the
nitrogen oxides cease at once.

The analysis of mixtures of nitrogen oxides can be effected by examination of their ultra-red absorption spectra;
all five oxides, and also ozone, can be detected by their
different maxima of absorption. In this way the nature of
the products obtained under various conditions of electrical
discharge, etc., can be conveniently examined[2].

Nitrites. In 1906 a method for the production of alkali
nitrites, by passing a mixture of air or oxygen and ammonia
over heated metallic oxides, was patented[3]. With ferric
oxide at 700° a continuous current of nitrous gas is obtained,
which is converted into sodium nitrite by absorption in
sodium hydroxide. Nitrites can also be prepared by oxidising atmospheric nitrogen to the appropriate extent in
the electric arc and maintaining the gases at a temperature
not below 300° until absorption in solution of alkali hydroxide has been effected: in this way the formation of
higher oxides is almost entirely prevented[4].

Chlorides. The composition of the chloride of nitrogen
obtained by the action of chlorine on ammonia has long
been uncertain: many analyses having shown that it contained hydrogen. In 1909 an analysis of the substance,

[1] A. Koenig and E. Elöd, *Ber.* 1914, **47**, 516.
[2] E. Warburg and G. Leithäuser, *Sitzungsber. K. Akad. Wiss.
Berlin,* 1908, 148.
[3] *D. R.-P.* 168272.   [4] *Ibid.* 188188.

purified by solution in carbon tetrachloride[1], showed that
there was less than one atom of hydrogen to each hundred
atoms of nitrogen. The composition of the substance is
evidently represented by the formula NCl_3. Two years previously another chloro-derivative of ammonia 'had been
prepared in a pure state[2]. Equimolecular quantities of
ammonia and sodium hypochlorite react according to the
equation :

$$NaOCl + NH_3 = NH_2Cl + NaOH.$$

By distilling in a vacuum, a faintly yellow, unstable oil is
obtained. This substance, to which the author gives the
name of chloroamine, reacts vigorously with alkalis, giving
ammonia and nitrogen : black nitrogen iodide is precipitated
from a solution of potassium iodide, whilst with ammonia a
small quantity of hydrazine is produced.

Nitrosyl perchlorate[3], $NO . ClO_4 . H_2O$, has been prepared
by passing oxides of nitrogen, prepared by the action of
nitric acid on sodium nitrite, into perchloric acid. The salt
is obtained in colourless, doubly refracting leaflets, which
are slightly hygroscopic. Water produces decomposition,
resulting in a green solution. Alcohol, ether, acetone, and
primary aromatic amines ignite when mixed with the new
perchlorate, producing violent explosions.

Sulphides. When sulphur tetrachloride and ammonia
react, nitrogen sulphide is produced in accordance with the
equation[4] :

$$12SCl_4 + 16NH_3 = 3N_4S_4 + 48HCl + 2N_2.$$

The substance is also produced when sulphur is dissolved
in liquid ammonia. The action, which results in the formation of hydrogen sulphide, is a reversible one. The removal
of the hydrogen sulphide by means of silver iodide dissolved
in ammonia favours the formation of the nitrogen sulphide,

[1] D. L. Chapman and L. Voddon, *Trans. Chem. Soc.* 1909, **95,** 138.
[2] F. Raschig, *Verh. Ges. deut. Naturforsch. Arzte,* 1907, II, i, 120.
[3] K. A. Hofmann and A. Zedtwitz, *Ber.* 1909, **42**, 2031.
[4] Ruff and Geisel, *Ber.* 1904, **37**, 1573 ; *ibid.* 1905, **38**, 2659.

which can be obtained by evaporating off the ammonia from the filtrate left after the removal of the silver sulphide. Nitrogen sulphide is attacked by dry hydrogen chloride, yielding ammonia, but no nitrogen.

PHOSPHORUS

Pure ordinary phosphorus may be obtained by heating ordinary phosphorus with chromic acid solution, washing, drying in a vacuum at 80° C. and finally distilling[1]. The substance thus obtained is perfectly white, and remains so if kept in the dark in evacuated tubes, but it soon becomes yellow on exposure to light. If the phosphorus thus obtained be melted and subjected to fractional crystallisation in an exhausted vessel, a perfectly colourless and coarsely crystalline form is obtained. The substance melts sharply at 44° C.[2]

In 1914 a new form of white phosphorus, which changes reversibly into the ordinary modification, was obtained[3]. It was first produced by increasing the pressure on ordinary phosphorus to about 11,000 kilogram/cm.[2] at 60°. The transition temperature is a linear function of the pressure, and lies between − 76·9° at a pressure of 1 kilogram/cm.[2] and 64·4° at a pressure of 12,000 kilogram/cm.[2] By crystallisation from carbon disulphide at low temperatures, the new modification was obtained in the form of microscopic crystals belonging to the hexagonal system.

Experiments on phosphorus vapour show that between 500° and 700° and 240 and 300 mm. pressure it obeys Boyle's and Gay-Lussac's laws and that it consists entirely of P_4 molecules[4]. At higher temperatures dissociation into P_2

[1] J. Böeseken, *Proc. K. Akad. Wetensch. Amsterdam*, 1907, **9**, 1613; *ibid., Rec. Trav. Chim.* 1907, **26**, 289.

[2] A. Smits and H. L. de Leeuw, *Proc. K. Akad. Wetensch. Amsterdam*, 1911, **13**, 822.

[3] P. W. Bridgman, *J. Amer. Chem. Soc.* 1914, **36**, 1344.

[4] Stock, Gibson and Stamm, *Ber.* 1912, **45**, 3527.

sets in, especially at lower pressures. At 1200° and 175 mm. pressure two-thirds of the P_4 had dissociated. The only equilibrium concerned is $P_4 \rightleftharpoons 2P_2$.

The phosphorescence associated with the slow oxidation of phosphorus appears to be due not to the primary oxidation of the element but to subsequent oxidation of the phosphorous oxide produced. It has been shown that if air is passed rapidly over phosphorus the phosphorescence is drawn in the direction of the current, and on further increasing the speed it becomes completely detached from the phosphorus: the dark region may be several metres in length. In this dark region there is neither production of ozone nor ionisation, so that it is probable that all three phenomena are directly connected. Confirmation of this is found in the fact that phosphorus does not glow in pure oxygen at ordinary temperatures[1].

Red Phosphorus. Pure red phosphorus has been obtained by heating phosphorus tribromide with mercury in a sealed tube at 100°—170° for several days[2]. The product is a cinnabar red micro-crystalline powder, which becomes brownish-black at 250° but regains its colour on cooling. It is insoluble in carbon disulphide, oxidises slowly in moist air and inflames in air at approximately 300°. Much discussion has taken place as to the relation between red phosphorus and Hittorf's variety, but no conclusive evidence has been brought forward to disprove Chapman's[3] conclusion arrived at in 1899 that the two varieties are identical. A good deal of the work on the subject has been vitiated owing to the fact that commercial red phosphorus, which always contains yellow phosphorus, was used. There is evidence that pure red phosphorus undergoes a slow change to ordinary phosphorus, and this fact may account for some of the discrepancies in the work.

---

[1] See also oxides of phosphorus.
[2] L. Wolf, *Ber.* 1915, **48**, 1272.
[3] *Trans. Chem. Soc.* 1899, **75**, 734.

Hittorf's Phosphorus. This is obtained by heating lead and phosphorus in a sealed tube at about 500° for eighteen hours and allowing the phosphorus to crystallise from the molten lead. The substance may contain half its weight of lead, and the purest product obtainable by treating first with dilute and then with concentrated nitric acid still contains 1·5 per cent. of lead[1].

Black Phosphorus. This is prepared[2] by heating ordinary phosphorus to 200° under a pressure of 12,000 kilogram/cm.[2] During the first fifteen minutes there is little change, then transition sets in and the pressure drops by more than 8000 kilograms. The density of the black modification so obtained is high, being about 2·69 against 1·83 for ordinary white phosphorus and 2·34 for Hittorf's red phosphorus. This variety is characterised by the fact that it can be heated to 400° in air without catching fire; it can be ignited with difficulty. Like red phosphorus it is attacked by cold nitric acid and is not dissolved by carbon disulphide. It is not found possible to convert red phosphorus into black phosphorus.

Violet Phosphorus. This modification is obtained by adding a trace of sodium (to act as catalyst) to some ordinary white phosphorus and then subjecting the latter to a pressure of 4000 kilogram/cm.[2] at the ordinary temperature[3]. The substance is then heated to 200° keeping the volume constant. The pressure is then raised to 12,500 kilogram/cm.[2] for twenty minutes and then to 130,000 kilogram/cm.[2] for forty-five minutes. The apparatus is then cooled and the pressure released at the ordinary temperature. The white phosphorus is entirely transformed into violet phosphorus, the density of which is 2·35. If the experiment is carried out with a mixture of white and red phosphorus the former only is converted, the latter remaining unchanged.

[1] Stock and Gomolka, *Ber.* 1909, **42**, 4510.
[2] P. W. Bridgman, *J. Amer. Chem. Soc.* 1914, **36**, 1344.
[3] *Ibid.* 1916, **38**, 609.

Oxides. Pure oxygen at atmospheric pressure oxidises phosphorus directly to phosphoric oxide only, but if the concentration be diminished by lowering the pressure or by mixing with inert gas then phosphorous oxide also is produced[1]. The process can be so regulated as to produce crystals of phosphorous oxide. This is exceedingly inflammable and on admission of air or oxygen takes fire before any phosphorus which may be present. The monoxide of phosphorus originally prepared in 1897 by the interaction of phosphoryl bromide and hydrogen phosphide[2], and upon the existence of which doubts have been cast, has recently been prepared by submitting a mixture of hydrogen and phosphoryl chloride to the silent discharge[3].

Hydrides. In 1909 two new solid hydrides of phosphorus were prepared[4]. Calcium phosphide, on treatment with water, gives a mixture which, on passing over granular calcium chloride, gives a yellow deposit. On treatment with cold dilute hydrochloric acid, the calcium is dissolved, and the solid left is washed with water and then with alcohol and ether. The hydride is a canary-yellow powder, which, when first prepared, has no odour. On standing in the air, it rapidly changes, especially in sunlight, giving off a spontaneously inflammable phosphine. It is remarkably insoluble in reagents, and on analysis it is stated to have the composition $P_{12}H_6$. On heating in a vacuum, it turns red and evolves pure phosphine, the residue being the second new hydride,

$$5P_{12}H_6 = 6P_9H_2 + 6PH_3.$$

The second hydride is stable in dry air, but in presence of moisture it is converted into phosphine and phosphoric acid.

[1] E. Jungfleisch, *Compt. rend.* 1907, **145**, 325.
[2] Besson, *Compt. rend.* 1897, **125**, 1032.
[3] Besson and Fournier, *Compt. rend.* 1910, **151**, 876.
[4] A. Stock, W. Böttcher, and W. Lenger, *Ber.* 1909, **42**, 2839.

Chlorides. A new chloride has been prepared by submitting a mixture of phosphorus trichloride and hydrogen to the action of the silent electric discharge[1]. A yellow solid is formed at the same time. The liquid is fractionally distilled under diminished pressure, and after purification is found to have the composition P_2Cl_4. It is easily oxidisable in air, taking fire spontaneously in certain circumstances. With water, phosphorous acid and a yellow solid are produced.

Sulphides. The pentasulphide[2] has been prepared by heating a solution of phosphorus and sulphur with a trace of iodine in carbon disulphide at a temperature of 120°—130°. The substance in appearance resembles flowers of sulphur. It is not very soluble in carbon disulphide and has a molecular formula P_4S_{10}.

The existence of only three sulphides has been proved, namely P_4S_3, P_4S_7, and P_4S_{10}, other so-called sulphides being merely mixtures[3].

Acids. The ionisation constants of the first two hydrogen ions of pyrophosphoric acid are almost identical, and very near that of the first ion of orthophosphoric acid[4]. Those of the third and fourth ions of pyrophosphoric acid are near to one another but widely different from the first. This suggests that the acid possesses an unsymmetrical structure. The most probable structural formulæ for ortho- and pyrophosphoric acids are :

[1] Besson and Fournier, *Compt. rend.* 1910, **150**, 112.
[2] Stock and Thiel, *Ber.* 1905, **38**, 2719.
[3] Stock, *Ber.* 1908, **41**, 558, 657.
[4] D. Balareff, *Zeitsch. anorg. Chem.* 1914, **88**, 133.

TANTALUM

In 1905 the decomposition of tantalum tetroxide enabled the metal to be obtained in the form of filaments suitable for the construction of incandescent electric lamps[1]. The fact that the metal when white hot does not combine with oxygen if the pressure be below 20 mm. gives a clue to the explanation of the method of preparing the metal, which is effected by heating electrically in a vacuum rods made of the compressed oxides. At the high temperature so obtained the oxide is decomposed and the oxygen removed by continued exhaustion. The metal melts about 2250° and combines extraordinary hardness with great ductility. Vanadium and columbium may be obtained in a similar manner from their oxides.

ANTIMONY

Antimony hydride is best prepared by the action of hydrochloric acid upon an alloy of magnesium and antimony containing 33 per cent. of the latter[2]. The gas dissolves in ether, ligroin and benzene but the solution quickly becomes turbid ; the best solvent is carbon disulphide which dissolves 250 times its own volume. It is readily decomposed by electric discharges and at times decomposes spontaneously into antimony and hydrogen. The dry gas is stable, the presence of moisture promotes its instability, and it is attacked by the air at the ordinary temperature. It reacts readily with oxygen forming antimony and water, and converts nitric oxide into nitrous oxide, nitrogen, and ammonia, antimony being formed at the same time. The halogens decompose it readily. By the action of air and oxygen on the liquid at − 90°, an unstable yellow modification of antimony is formed, which dissolves in carbon disulphide, forming an intensely yellow solution. At − 50°, this form passes into the metallic variety of antimony.

[1] von Bolton, *Zeit. Elektrochem.* 1905, **11**, 45, 503.
[2] Stock and Guttmann, *Ber.* 1904, **37**, 885.

PERIODIC GROUP VI

OXYGEN

ALTHOUGH liquid nitrogen and liquid hydrogen can be easily converted into the solid state by evaporation at diminished pressure yet it was not until 1911 that the conversion with oxygen was effected. Sir James Dewar[1] then succeeded in obtaining solid oxygen by evaporating the liquid at the very low pressure obtainable by his cooled charcoal method. The melting-point pressure was found to be as low as $1 \cdot 11$ mm., and even when this was attained solidification did not at once take place owing to super-cooling in the liquid oxygen.

The facts of auto-oxidation may be explained by assuming that an unsaturated molecule is added to a molecule of oxygen one atom of which functions as a quadrivalent atom[2]. The grouping $O : O :$ is probably more common and more stable than that of two quadrivalent oxygen atoms, just as in the case of nitrogen the grouping $: N : N \cdot$ is more common and stable than two quinquevalent nitrogen atoms. Under this assumption hydrogen peroxide would be regarded as $O : OH_2$ existing in the unimolecular state only, whereas water in the bimolecular form would be $H_2O : OH_2$. Other formulæ proposed are: ozone $O : O : O$; barium peroxide $O : OBa$ and sodium hydrogen peroxide $O : OHNa$.

Ozone. The ordinary ozone tube used in the laboratory rarely gives any large percentage of ozone, 3 to 8 per cent. being the usual yield. Oxygen containing a much greater percentage of ozone can be prepared by electrolysis of dilute sulphuric acid by using an anode that exposes very

[1] *Proc. Roy. Soc.* 1911, [A] **85**, 589.
[2] Julius Meyer, *J. pr. Chem.* 1905, (ii), **72**, 278.

little surface to the liquid[1]. The most effective anode is made by embedding platinum foil in glass and grinding away the edge so that only a line of 0·1 mm. breadth is exposed. The proportion of ozone in the oxygen evolved amounts to as much as 23 per cent. Ozone is said to be produced[2] when air is blown over a glowing Nernst filament, and since the stability of endothermic compounds is greater the higher the temperature the conversion should be possible. The reported production of ozone at high temperatures has been attributed to the formation of nitric oxide[3].

Ozone may be prepared by chemical means by gently heating a mixture of ammonium persulphate and nitric acid in an atmosphere of carbon dioxide[4]. The evolved gases are purified by passing them through dilute alkali and are found to contain 3—4 per cent. ozone, the rest being mainly oxygen.

When ozone is decomposed at 350° a phosphorescent light is observed[5] which is more marked if a hot glass rod is brought near the surface of liquid ozone. It was shown[6] in 1910 that when ozone decomposes in presence of carbon monoxide the dioxide is produced. The mixture of gases reacts strongly in sunlight and slowly in the dark.

Oxides. The consideration of cases of reduction of metallic oxides by mere raising of the temperature is generally confined to those in which reduction takes place from a higher to a lower oxide only, the case of mercury being treated as exceptional with regard to the reversibility of the action between metal and oxygen. Experimental investigations have shown that many "stable"

[1] F. Fischer and K. Bendixsohn, *Zeitsch. anorg. Chem.* 1909, **61**, 13.
[2] F. Fischer and Hans Marx, *Ber.* 1907, **40**, 443, 1111.
[3] Clement, *Ann. Physik,* 1904, [IV], **14**, 334.
[4] Malaquin, *J. Pharm. Chim.* 1911, [VII], **3**, 329.
[5] M. Beger, *Zeitsch. Elektrochem.* 1910, **16**, 76.
[6] Clausmann, *Compt. rend.* 1910, **150**, 1332.

oxides can easily undergo dissociation at not very high temperatures if the pressure or concentration of oxygen be kept sufficiently low. For example[1], in the cathode light vacuum cadmium oxide at 1000° dissociates into metal and oxygen, lead oxide does so at 750°, and bismuth oxide at a still lower temperature ; some sulphides also dissociate, giving sulphur and metal or lower sulphide. For a number of oxides the dissociation pressures at various temperatures have been measured[2], and where possible also the temperatures have been ascertained at which the pressure reaches that of oxygen in the air, this being, of course, the temperature at which decomposition would set in under ordinary conditions. One example may be cited : In air, cupric oxide would begin to decompose in the neighbourhood of 1025° with formation of oxygen and cuprous oxide ; but this is a much more stable compound, for even at that temperature its dissociation pressure is not more than 1 mm. Other determinations of a more or less similar kind have been made regarding the oxides of chromium (alone, and in presence of copper oxides), manganese, iron, and iridium.

In view of the dissociation phenomena just referred to, the mechanism of the reduction of oxides by means of, say, carbon would appear to be uncertain ; is the oxygen directly attacked by the reducing agent, or does this merely act as a kind of absorbent for the oxygen liberated by dissociation ? This problem also has been investigated[3], and it would appear that, in many cases at least, carbon acts directly on the oxide ; evolution of gas takes place at temperatures far below those at which direct dissociation can be detected in absence of carbon. It has been satisfactorily proved that even magnesium oxide can be reduced by carbon at a temperature of 1700° ; this had

---

[1] Damm and Krafft, *Ber.* 1907, **40**, 4775.
[2] H. W. Foote and E. K. Smith, *J. Amer. Chem. Soc.* 1908, **30**, 1344.
[3] H. C. Greenwood, *Trans. Chem. Soc.* 1908, **93**, 1483.

been surmised from observed facts[1], but in 1908 was demonstrated[2] very conclusively by dissolving the resulting magnesium vapour in metallic copper (which itself has no action on magnesia), and, still more conclusively, by condensing it as a metallic mirror on the walls of an evacuated vessel.

The constitution to be assigned to the members of the two groups of higher oxides, commonly designated as peroxides and exemplified by MnO_2 and BaO_2, are expressed by the general formula $M\underset{O}{\overset{O}{\big\langle}}$ and $M\underset{O}{\overset{O}{\big\langle}}\,\big|$; the former are oxides of the ordinary type, possessing feebly basic properties (capable of yielding an iron $M^{....}$) or feebly acidic properties or possibly both, while the latter are salts of hydrogen peroxide. The name *peroxidate* given to the latter is in many respects convenient.

SULPHUR

Allotropic Forms. In addition to the rhombic and prismatic crystalline forms of sulphur a third form, called nacreous sulphur, exists. This may be prepared in lustrous, needleshaped crystals by heating sulphur to 150°, cooling to 98° and making it crystallise by scratching the containing tube with a glass rod. Colloidal sulphur can be obtained from the sulphur which separates when sodium thiosulphate solution is added to cold concentrated sulphuric acid[3]. The liquid is somewhat diluted, heated, filtered through glass wool, and allowed to cool ; these processes are repeated so 'ong as any sulphur, which will not redissolve, is precipiated. A cloudy, yellowish-coloured mass is thus obtained,

[1] Lebeau, *Compt. rend.* 1907, **144**, 799.

[2] R. E. Slade, *Trans. Chem. Soc.* 1908, **93**, 327.

[3] M. Raffo, *Zeitsch. Chem. Ind. Kolloide*, 1908, **2**, 358.

which, on warming, forms a liquid, which is apparently perfectly clear ; it separates again on cooling. This mass is separated as completely as possible by means of a centrifuge, and is partly washed in a similar manner, after which the remaining acid is neutralised with sodium carbonate. The product is soluble in water, but if the attempt is made to remove the sodium sulphate from the hydrosol by dialysis, insoluble sulphur is very soon deposited. Dilute solutions, in which the sulphuric acid has not been neutralised, can be preserved for a long time, but the hydrosol very easily precipitates insoluble sulphur on addition of various salts.

The peculiarities exhibited by molten sulphur and the differences observed in different varieties of precipitated sulphur are now satisfactorily explained as a result of the investigations of A. Smith[1] and his collaborators. When sulphur is fused, two modifications, which are distinguished as Sλ and Sμ, result and these are dynamic isomerides. Sλ is comparatively mobile, is light in colour and is the more stable form below 160° ; Sμ is viscous and dark ; the two varieties are only partially miscible. All liquid sulphur contains both of these modifications up to high temperatures, the quantity of each present depending ultimately on the temperature only as in ordinary transformations of liquid dynamic isomerides. Up to 160° the equilibrium quantity of Sμ is completely dissolved by Sλ. At this temperature there are probably two phases, a saturated solution of Sμ in Sλ and a saturated solution of Sλ in Sμ. Above 160° only the latter solution is stable and the amount of Sλ diminishes as the temperature rises. If the temperature is raised rapidly there may be a decided lag so that some time may elapse before equilibrium is established. The rate of transformation can be greatly retarded by certain agents, such as sulphur dioxide or hydrochloric acid, but

---

[1] *J. Amer. Chem. Soc.* 1905, **27**, 797, 979; *Zeit. physikal. Chem.* 1905, **52**, 602; 1906, **54**, 257.

68   SULPHUR

the effect can be reversed by others, such as ammonia[1].
Advantage is taken of this in determining the state of
equilibrium at any given temperature ; equilibrium is
established rapidly with the help of ammonia, then a retarding gas is passed in and the liquid sulphur rapidly
chilled in water. The solid obtained from $S\lambda$ is crystalline
and soluble in carbon disulphide, while that from $S\mu$ is
amorphous and insoluble. When ordinary sulphur is heated
there is always considerable retardation owing to the in-
variable presence of sulphur dioxide in sulphur which has
been exposed to air[2]. When sulphur is absolutely pure no
amorphous sulphur is obtained on rapidly cooling liquid
sulphur from its boiling-point.

The "natural freezing-point[3]" of the mixture of $S\lambda$ and
$S\mu$ is 114·5° and at this temperature the mixture contains
3·7 per cent. of $S\mu$. This freezing-point varies with the
nature of the solid separating and with the proportion of
$S\mu$ in the liquid. The "ideal freezing-points" when no $S\mu$
is present and the "natural freezing-points" when an
equilibrium proportion of $S\mu$ is present have been obtained
for three varieties of the element as follows :

	Ideal F.P.	Natural F.P.
Rhombic sulphur S_1 or $S\alpha$	119·3°	114·5° (3·6 °/₀ $S\mu$)
Prismatic sulphur S_2 or $S\beta$	112·8°	110·2° (3·4 °/₀ $S\mu$)
Nacreous sulphur S_3	106·8°	103·4° (3·1 °/₀ $S\mu$)

In 1912 the discovery of yet another allotropic form was
announced[4]. If a solution of sulphur in sulphur chloride
saturated at the ordinary temperature is heated to 150°
and then cooled it dissolves a further and considerable
quantity of sulphur. This cannot be explained by the
formation of $S\mu$, since $S\mu$ at the ordinary temperature is

---

[1] A. Smith and C. M. Carson, *Proc. Rôy. Soc. Edin.* 1906, **26**, 352;
J. Amer. Chem. Soc. 1907, **29**, 499.
[2] See under oxides of sulphur.
[3] A. Smith and C. M. Carson, *Zeitsch. physikal. Chem.* 1911, **77**, 661.
[4] A. H. W. Aten, *Zeitsch. physikal. Chem.* 1912, **81**, 257 ; 1913, **83**, 442.

ónly sparingly soluble in sulphur chloride and separates
out quite easily on cooling a hot solution. Also when
sulphur is heated to 125° and rapidly cooled it becomes
more soluble in sulphur chloride ; again heated sulphur is
more soluble in toluene than unheated sulphur and the
solubility is greater the more solid sulphur there is present.
The results point to a new modification of sulphur which
is named Sπ. Further investigation shows that when sulphur is heated to 170° and rapidly cooled it contains just
as much Sπ as when it is heated to 445° and rapidly cooled.
The maximum amount of Sπ which it has been possible to
obtain by heating sulphur to various temperatures is about
6·5 per cent. after the sulphur has been heated to 180°.
All solutions containing Sπ are deep yellow ; for example,
a carbon disulphide solution containing 18 atoms per cent.
has a colour analogous to that of a concentrated aqueous
solution of potassium chromàte. On cooling such a solution
to − 80°, the whole of the Sλ separates out, leaving only
Sπ in solution. Attempts were made to obtain solid Sπ
by evaporation of such solutions but at first only Sμ was
obtained. It was found, however, that if the solution were
evaporated in a vacuum at a temperature of − 80°, the
residue left was almost entirely soluble in toluene, a very
small quantity of Sμ being left. It is extremely unlikely
that solutions of Sπ are really solutions of Sμ. Confirmatory
evidence of this was found in the action of light on the
solutions. Two solutions, one containing Sλ and the other
containing Sλ and Sπ, were exposed to a strong light ; in
each case Sμ was immediately precipitated, but much less
was obtained in the case of the solution containing both
Sλ and Sπ than in the case of the solution containing only Sλ.
Clearly, if Sλ and Sμ were identical, more Sμ would have
been precipitated in the case of the solution containing Sλ
and Sπ. Some of the properties of sulphur such as the
change in viscosity with temperature and the variation in
crystallisation power with previous thermal treatment

can be better explained when the presence of Sπ is taken into account[1].

A fourth modification Sϕ is obtained by adding concentrated hydrochloric acid at 0° to a cold solution of sodium thiosulphate and shaking the liquid with toluene[2]. After a short time Sϕ separates out from the toluene as orange-yellow crystals. As regards precipitated sulphur, the conclusions drawn are that the sulphur separates first in liquid form as Sμ, which, however, changes more or less rapidly to Sλ, and ultimately to Sα. The rate of transformation varies greatly with the nature of the liquid in contact with the liberated sulphur. The soluble varieties of precipitated sulphur are merely Sα; the insoluble varieties are those in which solidification has taken place before any great proportion of Sμ has been transformed into Sλ.

Hydrogen Sulphides. A convenient method of preparing hydrogen sulphide[3] is by dropping water on solid aluminium sulphide; the latter is prepared by igniting, by means of a magnesium ribbon fuse, a mixture of sulphur and aluminium in a crucible.

From time to time conflicting statements have been made as to the true composition of "hydrogen persulphide." After first having been determined to be somewhat complex, it was later assumed to be represented by the formula H_2S_2, from analogy to hydrogen peroxide; then various higher sulphides were supposed to exist, corresponding with the polysulphides of the metals, but this was contradicted, and it was stated that whatever polysulphide might be added to an acid, the resulting hydrogen compounds were H_2S and H_2S_5 only[4]. This in turn is now shown to be incorrect, on the authority of independent investigators. By fractional distillation (under low pressure) of the oil ob-

---

[1] A. H. W. Aten, *Zeitsch. physikal. Chem.* 1913, **86**, 1.

[2] *Ibid.* 1914, **88**, 321.

[3] H. Fonzes-Diacon, *Bull. Soc. Chim.* 1907 (IV), **1**, 36.

[4] Rebs, *Annalen*, 1888, **246**, 356.

tained by pouring alkali polysulphide solution into hydrochloric acid, the disulphide H_2S_2[1], and the trisulphide H_2S_3[2],
have been isolated as unstable liquids. The composition has
been exactly determined by improved methods of analysis
(the hydrogen being driven out as hydrogen sulphide, which
could be accurately estimated), and the molecular weight
by cryoscopic methods. Evidence is also published for the
existence of compounds from H_2S_5 to H_2S_9[3]. The result of
the consideration of the constitution of these polysulphides
is not unfavourable to Mendeléeff's conception of a possible
"homologous series" formed from HSH by successive re-
placements of H by SH; the general formula for such a
series would be of course H_2S_n[4].

In view of various investigations carried out in recent
years, it has been suggested that tetrathionic acid has a
peroxidic constitution, $(HO_2S_2).O.O.(S_2O_2H_2)$, and not the
persulphidic constitution, $(HO_3S).S.S.(SO_3H)$, generally
assumed, the reason being that alkaline reducing agents
apparently abstract oxygen directly from its salts; thus,
alkaline arsenite solution forms arsenate as well as mono-
thioarsenate. It has been shown[5], however, that there are
serious objections to the assumption of a peroxide union in
the tetrathionates, and that Mendeléeff's persulphide formula, which fits in so well with the general behaviour of the
salts, can perfectly well account for these new facts also.

Oxides. When sulphur is burnt in air 7 per cent. of it
is converted into the trioxide, while when burnt in oxygen
only about 2·7 per cent. is so converted[6]. Two explanations
of this have been given: one that the higher yields with air
are due to the formation of oxides of nitrogen which act

[1] J. Bloch and F. Höhn, *Ber.* 1908, 41, 1961, 1975.
[2] *Ibid.* 1971; R. Schenck and V. Falcke, *ibid.* 2600.
[3] G. Bruni and A. Borgo, *Atti R. Accad. Lincei*, 1907 (v), 16, ii, 745.
[4] J. Bloch, *Ber.* 1908, 41, 1980.
[5] T. S. Price and D. F. Twiss, *Trans. Chem. Soc.* 1907, 91, 2021;
J. E. Mackenzie and H. Marshall, *Trans. Chem. Soc.* 1908, 93, 1726.
[6] J. H. Kastle and J. S. McHargue, *Amer. Chem. J.* 1907, 38, 465.

as carriers, while the other regards the higher temperature
produced in oxygen as being less favourable to the formation of the trioxide.

Sulphur is oxidised in air at the ordinary temperature,
especially in sunlight[1]. The production of sulphur dioxide
in this way sufficiently explains the antiseptic properties
of the element without the suggestion that they are due to
hydrogen peroxide produced by plants.

Metallic sulphides are attacked by steam, at temperatures varying from incipient redness to a white heat[2]; sulphur dioxide, hydrogen, sulphur, and in some cases the
metals themselves are produced. The presence of sulphur
dioxide in volcanic gases may be due to the action of steam
on metallic sulphides.

Sulphuric Acid. In 1911 a theory of the chamber sulphuric acid process was put forward[3] in which it was assumed that an unstable nitrososulphonic acid was first
formed by the action of sulphur dioxide on nitrous acid as
shown by the equation:

$$HNO_2 + SO_2 = ONSO_3H,$$

and that this acid is at once transformed by more nitrous
acid into nitrosisulphonic acid:

$$ONSO_3H + HNO_2 = H_2NSO_5 + NO,$$

and that this acid then decomposes into sulphuric acid and
nitric oxide:

$$H_2NSO_5 = H_2SO_4 + NO.$$

The nitrosisulphonic acid was stated to be identical with
Sabatier's violet acid.

Experiments[4] have been made with a view to testing
this theory, and it was found that the only gaseous product

[1] Harpf, *Zeit. anorg. Chem.* 1904, **38**, 142.
[2] Gautier, *Compt. rend.* 1906, **142**, 1465; *ibid.* **143**, 7.
[3] Raschig, *J. Soc. Chem. Ind.* 1911, **30**, 166.
[4] Divers, *J. Soc. Chem. Ind.* 1911, **30**, 594; Reynolds and Taylor, *ibid.* 1912, **31**, 367.

of the interaction of sulphurous and nitrous acid is nitrous oxide, and chamber crystals cannot exist even in 60 per cent. sulphuric acid, while Sabatier's violet acid is only formed in solutions containing chamber crystals. These experiments showed not only that the above theory is wrong but also indicate the futility of all attempts to account for the formation of the bulk of the sulphuric acid in the chamber by the intermediary formation of nitrogen sulphonic acids. It must be regarded as doubtful if nitrososulphuric acid (chamber crystals) play an important part in the process. The experimenters also came to the conclusion that the old views substantially represent the course of the process, namely :

(1) $SO_2 + H_2O = H_2SO_3$;

(2) $H_2SO_3 + NO_2 = H_2SO_4 + NO$;

(3) $2NO + O_2 = 2NO_2$.

Caro's Acid. Though this acid was discovered in 1898 its composition was not finally settled till some ten years later. Two formulæ had been assigned to the substance, H_2SO_5[1] and $H_2S_2O_9$[2]. The analysis of its salts is inconclusive since a salt $MHSO_5$ is indistinguishable from $M_2S_2O_9 . H_2O$. Experiments[3] made with the acyl and benzoyl derivatives show that the acid is monobasic and that therefore its potassium salt has the formula $KHSO_5$. The following graphical formulæ show the relation between this acid and hydrogen peroxide on the one hand and sulphur heptoxide on the other:

O—H	O—SO₂—OH	O—SO₂—OH	O—SO₂\
|	|	|	| >O
O—H	O—H	O—SO₂—OH	O—SO₂/
Hydrogen peroxide	Caro's acid	Persulphuric acid	Sulphur heptoxide

The reaction by which this acid was first prepared is reversible, $H_2O_2 + H_2SO_4 \rightleftarrows H_2SO_5 + H_2O$, and so the product could

[1] Baeyer and Villiger, *Ber.* 1901, **34**, 853.

[2] Armstrong and Lowry, *Proc. Roy. Soc.* 1902, **70**, 94.

[3] R. Willstätter and E. Hauenstein, *Ber.* 1909, **42**, 1839.

not be obtained pure. An acid of 92·5 per cent purity can however be obtained by the action of 100 per cent. hydrogen peroxide on sulphur trioxide[1].

SELENIUM

The value of photoelectric cells in physical and astrophysical research is now fully recognised, and as selenium is the basis of these cells attempts have been made to prepare the element in a more sensitive form. If vitreous selenium, melted at 200°, is cooled rapidly under pressure it is transformed into a new variety, violet-grey in colour[2]. This consists of slender crystals which are very sensitive photoelectrically, but very unstable. By using a solid solution of this substance in vitreous selenium it is possible to prepare a very sensitive photometric cell.

[1] H. Ahrle, *J. pr. Chem.* 1909 [ii], **79**, 129; *Zeitsch. angew. Chem.* 1909, **22**, 1713.
[2] L. Angel, *Bull. Soc. Chim.* 1915 [iv], **17**, 10.

PERIODIC GROUP VII

FLUORINE

Hydrofluoric Acid. Many investigations[1] on the basicity of hydrofluoric acid have been made and both from conductivity and cryoscopic determinations it is concluded that the acid is dibasic and its molecule is represented by the formula H_2F_2. In addition to its use in etching glass, the acid readily removes rust from iron.

CHLORINE

A study of the freezing-point and boiling-point curves obtained for mixtures of the halogens has shown that chlorine and bromine do not unite to form any compound —gaseous, liquid or solid ; the "compounds" previously described are solutions and mixed crystals of the elements[2]. The compound ICl exists not only in the liquid state, but also in the state of vapour; at 100° it is very slightly dissociated. The results of the experiments also showed that chlorine and bromine have less mutual affinity than bromine and iodine.

Hypochlorite. Experiments[3] on the mechanism of the conversion of alkaline solutions of sodium hypochlorite into sodium chlorate show that at a temperature of 50° the reaction is of the second order rather than the third, and, therefore, the first stage is represented by the equation $2NaClO = NaClO_2 + NaCl$. There is also an evolution of

[1] R. Kremann and W. Decolle, *Monatsh.* 1907, **28**, 917; G. Pellini and L. Pegorado, *Zeitsch. Elektrochem.* 1907, **13**, 621.

[2] B. J. Karsten, *Zeitsch. anorg. Chem.* 1907, **53**, 365.

[3] F. Foerster and P. Dolch, *Zeitsch. Elektrochem.* 1917, **23**, 137.

oxygen, which evidently occurs according to the equation $2NaClO = O_2 + 2NaCl$. Further, the transformation of the chlorite into chlorate is a bimolecular reaction, and follows the equation $NaClO + NaClO_2 = NaClO_3 + NaCl$. The formation, therefore, of chlorate from hypochlorite occurs according to the equations (1) $2NaClO = NaClO_2 + NaCl$, and (2) $NaClO + NaClO_2 = NaClO_3 + NaCl$. The latter reaction proceeds much more rapidly than the former.

IODINE

Iodine dioxide, first obtained in 1844, has been prepared in a pure condition[1] by acting with concentrated sulphuric acid on iodic acid till iodine begins to be evolved, and then cooling the mixture. A yellow crystalline crust is obtained. This, after washing with water and then with ether, gives the pure dioxide of iodine. Owing to its insolubility and its decomposition on heating its molecular weight cannot be found. By the direct action of sulphur trioxide on iodine dioxide, a solid compound, $I_2O_4 . 3SO_3$, is found.

[1] M. M. Pattison Muir, *Trans. Chem. Soc.* 1909, **95**, 656.

PERIODIC GROUP VIII

IRON

MOISSAN[1] in 1906 succeeded in distilling iron and several other metals by heating them in an electric furnace and obtaining crystalline distillates. The metals nearly allied to iron decrease in volatility in the following order : manganese, nickel, chromium, iron, uranium and tungsten.

Passivity of Iron. The problem of the passivity of iron is one which has now for very many years engaged the attention of chemists, and not a few explanations of the phenomenon, both physical and chemical, have been advanced. In spite of a large amount of experimental work carried out in recent times, a final solution of the problem has perhaps not yet been reached, although the opinion which seems now to be most probable is that it is due to the formation of a layer of ferroso-ferric oxide. This opinion is, however, by no means unanimous.

The subject of the passivifying of iron, and of the activifying of passive iron, has been very fully investigated by Heathcote[2], who has studied, not only the chemical nature of solutions which bring about passivity and activity, but also the electrometric state of the metal in the two states. As passivity does not appear to be a static phenomenon, not a few of the conflicting results obtained by previous workers are attributed to differences in the degree of passivity. Iron is regarded as passive when, after plunging in nitric acid, of sp. gr. 1·2, shaking for a moment in the acid, and then holding motionless, no chemical action can

[1] Moissan, *Compt. rend.* 1906, **142**, 274.
[2] H. J. L. Heathcote, *J. Soc. Chem. Ind.* 1907, **26**, 899.

be detected at the surface by the unaided eye, the temperature of the acid being about 15—17°. Passivification can be produced, not only by immersion of the metal in certain solutions, but also by the electric current, and a close connection has been found to exist between the phenomena observed in the two cases. It is therefore believed that the process of passivification is in all cases electrolytic, and that, when no external current is employed, current is generated between one part of the surface and another. From this standpoint, the process of passivification is pictured in the following manner. When iron is immersed in a solution, some of the metal passes into solution, forming ferrous ions, or the ferrous ions in the metal pass into the solution from all parts of the surface. The solution pressure of the iron will not, however, be the same at all points of the surface, but will be greater, for example, at the ridges and points than at the hollows on the surface. There will therefore arise local currents, and, if the solution will act as a depolariser at the cathode of these local circuits, the current may persist until passivification at the anodes of the local circuits is effected. This process may require a shorter or longer time, according to the depolarising power of the solution at the cathode and its chemical action on the active iron. If the rate of action of the solution on the passive particle is small, and if the solution can depolarise, currents between the passive and active parts may passivify the active areas, and passivity may thus be established over the whole surface. Since, however, the resistance to an electric current from or to a point is infinite, so this point will probably never become really passive. Likewise, the bottoms of crevices or of pits on the surface will remain active. So long as these exist, they will give rise to currents, while the metal remains passive. Passive iron is soluble, but these active particles will generate current sufficient to maintain it in the passive condition so long as the liquid will continue to combine with the electrolytic products and prevent the activification of the passive part. If the depolarisation fails, the activity will spread from the active areas.

The phenomena of passivity are best explained on the assumption of the formation of a layer of magnetic oxide. The hydroxyl ion can be liberated even from solutions which contain no oxygenated anions, and the results obtained go to show that it is the hydroxyl ion which effects passivity, and that this state is due to the formation of a solid phase. This view of the cause of passivity is also borne out by E.M.F. measurements[1].

The Rusting of Iron. Very many attempts to find a solution of the problem of the cause of the rusting of iron have been made during the last fifteen years, and a study of the work done and the conclusions drawn show not only that exceptional experimental skill is required but that the evidence must be very carefully considered from every possible standpoint. Until comparatively recently the action was supposed to be one of direct oxidation. In 1871 Crace-Calvert attributed the action to the presence of carbonic acid, and in 1888 Crum Brown gave his support to this idea. He imagined that ferrous carbonate and hydrogen were the first products and that the hydrogen combined with the oxygen dissolved in the water while the ferrous carbonate was oxidised to rust by the same agency: the carbon dioxide liberated in the latter action would be then capable of recommencing the cycle of operations and so the rusting would proceed indefinitely. A few years before this, however, Traube, having shown that hydrogen peroxide was produced during the action of water on several metals, put forward the suggestion that this substance was also a precursor of the rusting of iron, although he was unable to detect its presence. In 1905 Dunstan made many experiments from which it appeared that iron, oxygen and

[1] F. Haber and W. Maitland, *Zeitsch. Elektrochem.* 1907, **13**, 309.

liquid water are alone essential to the phenomenon and
that the carbon dioxide of the air plays but a small part[1].
Analyses of rust produced in different experiments agree
satisfactorily with the formula $Fe_2O_2(OH)_2$. An explanation
of the process of rusting was advanced which assumed the
production of hydrogen peroxide. In the following year
Moody[2] finally disposed of the hydrogen peroxide hypothesis by showing that bright iron would go on catalytically
decomposing redistilled hydrogen peroxide for some weeks
without any rusting taking place. The hydrogen peroxide
must, in fact, have inhibited the action, since all the conditions for rapid rusting were present, the oxygen being
even in the nascent state. Using clean iron and carefully
purified water and air, Moody was able to keep the three
substances in contact without rusting taking place: if however a little carbon dioxide was allowed to pass into the
tube rusting at once took place. Objections have been
raised to Moody's experiments on the ground that in
purifying the surface of the iron he had used a dilute
solution of chromic acid and hence rendered the surface
passive. That this is not a vital objection is shown by the
fact that (1) the iron did rust as soon as a trace of carbon
dioxide was introduced, (2) it is impossible to make iron
passive to ordinary water by treatment with dilute chromic
acid.

In 1903 Whitney[3] propounded a theory of rusting according to which an acid is not necessary. Since the
purest water yet obtained is slightly dissociated, metallic
iron would tend to pass into solution as ferrous hydroxide.
Tilden supports this theory, supposing that either the impurities or even the harder parts of the metal form couples

[1] Dunstan, Jowett and Goulding, *Trans. Chem. Soc.* 1905, **87**, 1548;
Cribb and Arnaud, *Analyst*, 1905, **30**, 225; Lindet, *Compt. rend.*
1904, **139**, 859.
[2] *Trans. Chem. Soc.* 1906, **89**, 720.
[3] *J. Amer. Chem. Soc.* 1903, **25**, 394.

with the purer iron. This last suggestion is supported by the fact that in almost all cases rust forms on iron in spots which only slowly extend over the surface of the bright metal. It is thus seen that two different problems are really involved in the discussion: one the behaviour of pure iron in presence of pure water and oxygen, and the other the behaviour of the different kinds of iron and its alloys in ordinary use[1]. The results of a series of investigations on the former problem by Lambert and Thomson[2] were published in 1910. It is perhaps not generally recognised that work of this kind demands an amount of care and precaution very far beyond what would be thought necessary in, for instance, a good atomic-weight determination. Iron was obtained of a very high degree of purity by electrolysing purified ferric chloride, using iridium electrodes. The iron deposited was converted into ferric nitrate by means of very pure nitric acid, and this salt repeatedly crystallised from nitric acid. The salt was ignited in an iridium boat, and reduced to metal in hydrogen, obtained by the only method which gives the gas in anything like a pure condition, namely, the electrolysis of a solution of carefully purified barium hydroxide. The iron was contained in a silica test-tube, which was enclosed in a tube of Jena glass. Water was distilled from a dilute solution of barium hydroxide, without actual boiling, through a trap to prevent any possibility of the hydroxide being carried over. The experiment was so arranged that the water vapour which was ultimately condensed on to the iron was condensed only on silica. Oxygen, prepared by the electrolysis of barium hydroxide, which has been shown by many workers to be free from ozone and other impurities, was admitted to the tube containing the iron. In these circumstances the three substances, iron, water,

---

[1] J. N. Friend, *J. Iron and Steel Institute* 1908, **77**, i, 5; W. A. Tilden, *Trans. Chem. Soc.* 1908, **93**, 1356.
[2] *Trans. Chem. Soc.* 1910, **97**, 2426.

and oxygen, were found to be capable of contact without chemical action for some months. Moody's contention therefore is amply confirmed, but it must not be considered that the question is finally set at rest. In some of Lambert and Thomson's experiments, rusting undoubtedly took place, although carbon dioxide and other impurities were excluded.

In 1910 Dunstan and Hill[1] brought forward much new evidence which supports the electrolytic theory of Tilden[2]. According to this theory for rusting to take place iron must not be uniformly pure ; one part must be capable of becoming electro-negative to the other, and also the water must be sufficiently impure to be a conductor. If the iron is pure or the water is pure, no rusting can take place. Hence, pure iron in water containing carbon dioxide would not rust, neither would iron containing iron carbide in ideally pure water, because in neither case could an electric current be started. If impure iron is kept in N/10-sodium hydroxide solution, it will remain unrusted for an indefinite period. Here is a condition of things which should favour chemical action, namely, the impurities capable of becoming electronegative to the pure metal, and the couple in contact with a conducting solution. How then is the freedom from rust to be explained on the electrolytic theory ? It may be that the impurities are destroyed by the alkali, a supposition which is perhaps supported by the fact that a solution containing less than 0·13 per cent. of sodium carbonate will allow rusting to take place. A more likely hypothesis is suggested by the electrolytic theory. The surface of commercial iron may be regarded as made up of pure iron and impurities such as the sulphide, phosphide, or carbide of iron and free carbon. These are electronegative to pure iron, and would become the cathode when the metal was placed in an alkaline solution. A transient

[1] *Trans. Chem. Soc.* 1910, **99**, 1835.
[2] *Ibid.* 1908, **93**, 1356.

electrolysis would take place and either or both of two things might happen. The impurities might be coated with hydrogen or the pure metal might receive a superficial coating of oxide. In any case, electrolysis would stop, and the metal would become passive. Dunstan and Hill describe an interesting experiment in which iron is placed in a dilute sodium chloride solution containing a few drops of phenolphthalein. In absence of air nothing happened, but when air was admitted patches of pink colour appeared on the surface of the metal, rusting taking place on the parts which were not coloured. This is to be explained not by supposing "a difference of potential set up on the surface of the iron by the action of the air," but by assuming that the polarisation of the cathode is destroyed by oxidation, and that therefore a continuous current can flow and the chemical action proceed.

In 1912 Lambert[1] developed this electrical theory and brought forward considerable evidence in support of it. In the first place, some absolutely pure iron was prepared, and it was found that no trace of corrosion took place when it was kept in contact with either pure water and pure oxygen, or air and ordinary tap-water, for an apparently indefinite period. The explanation of this fact is to be sought in the greater homogeneity of the iron. If the iron be really homogeneous, and all parts have the same solution pressure, then sufficient difference of potential will not exist, and no rusting will take place. Furthermore, some pieces of this pure iron which have been exposed to water and air for some months without showing any corrosion were carefully dried ; some of the pieces were then placed in an agate mortar and pressed strongly with an agate pestle, whilst others were left untouched. All were then put in contact with the water and exposed to the air. Those pieces which had been pressed rapidly corroded, rust forming on those parts of the iron which had not been pressed, whilst the parts which had been

---

[1] *Trans. Chem. Soc.* 1912, **101**, 2056,

in actual contact with the polished agate remained quite
bright. Those pieces of iron which had not been subjected
to pressure showed no sign of corrosion. Evidently, therefore, the application of the pressure induced a change in the
iron, giving it a new solution pressure, so that electrical
action was set up when it was placed in contact with water.
An interesting fact was also noticed in that chemically pure
iron can be exposed for an apparently indefinite time to a
saturated solution of copper sulphate or copper nitrate without the slightest trace of copper being deposited. Copper is,
however, deposited on the iron if it is pressed in an agate
mortar before being put in the solution, or if pressed with
a quartz rod whilst under the solution. It is interesting
to note that all the samples of pure iron may not behave
exactly in the same way, even though prepared from the
same specimen of ferric nitrate. Some will show no corrosion after contact with water in air for an indefinite time,
whilst other specimens show corrosion in a few hours. The
difference in behaviour cannot be due to different chemical
composition, and must be due to physical differences produced by variation in tempering and rate of cooling in the
preparation. It would seem probable from these results
that the fundamental cause of the corrosion of iron is not
carbon dioxide or any other acid, but that the cause is
rather to be sought in the difference of solution pressure of
various parts of the iron, differences which may persist even
in the most highly purified form of the metal.

Ferric Sulphides. The existence of ferric trisulphide
has long been regarded as uncertain, but recent work[1] appears to show that the compound can be obtained. When
moist ferric hydroxide, or ferric hydroxide suspended in
water, is treated with hydrogen sulphide, it becomes black,
owing to the formation of ferric trisulphide, in accordance
with the equation $2Fe(OH)_3 + 3H_2S = Fe_2S_3 + 3H_2O$. In a

[1] V. Rodt, *Zeitsch. angew. Chem.* 1916, **29**, i, 422.

moist condition in the absence of air, or in the presence of excess of hydrogen sulphide, it is transformed into a mixture of the disulphide and sulphide, thus: $Fe_2S_3 = FeS_2 + FeS$. This change takes about a week at the ordinary temperature, but only a few hours at 60°. The mixture produced, being only partly soluble in dilute hydrochloric acid, may easily be distinguished from the original trisulphide, which is readily and completely soluble. When dried in a vacuum over phosphoric oxide, ferric trisulphide is very stable.

When exposed to the action of the air in presence of alkali, ferric trisulphide gradually becomes pale yellow, sulphur being deposited. The product resembles limonite in appearance, and when dried gives a fine, yellow powder containing a constant percentage of water, which is less than that corresponding with $Fe(OH)_3$.

When precipitated ferrous sulphide is boiled with flowers of sulphur, iron disulphide is formed. Ferric disulphide is also precipitated when a solution of sodium trisulphide is added slowly to a boiling solution of ferrous sulphate, sulphur being liberated at the same time.

Since ferric disulphide is the final product of the action of hydrogen sulphide on ferric hydroxide in the absence of alkali, the production of iron pyrites in nature can be explained.

NICKEL

Variable results are sometimes encountered in effecting the separation of the sulphides of nickel, cobalt, manganese, and zinc by treating the mixture with dilute hydrochloric acid. It appears that there are three different modifications of nickel sulphide[1], which differ in respect of their solubility in acids. The precipitates from a neutral solution of nickel chloride and soluble sulphides consist mainly of a modification which dissolves readily in acids. The less soluble

[1] A. Thiel and H. Gessner, *Zeitsch. anorg. Chem.* 1914, **86**, 1.

modification is formed by boiling with water, or by long contact with a cold solution containing nickel, or with cold dilute acetic acid. All the modifications have the same composition, NiS, and the most soluble is the most readily oxidised by air.

α-nickel sulphide is soluble in mineral acids down to 0·01 N, whilst β-nickel sulphide is fairly rapidly dissolved by 2N-hydrochloric acid, and the γ-modification is not appreciably dissolved except on addition of oxidising agents. The differences in solubility are too great to be explained by colloidal conditions, and must be due to polymerisation.

INDEX OF AUTHORS' NAMES

88 INDEX OF AUTHORS' NAMES

SUBJECT INDEX

Printed in the United States
By Bookmasters